Neurodiversity

Celebrating Differences in a Neurotypical World

Megan Dennis

ISBN: 978-1-77890-465-3
Imprint: Montecito Hot Springs

Contents

Introduction to Neurodiversity

Understanding Neurodiversity

Definition and Concept of Neurodiversity

Neurodiversity is a term that has gained recognition in recent years and has sparked a paradigm shift in how we perceive and understand neurological differences. It represents a departure from the traditional medical model of viewing neurodivergent individuals as having a deficit or disorder that needs to be fixed or normalized. Instead, neurodiversity celebrates the natural variation in human brain wiring and recognizes that neurological differences are simply a part of the natural diversity of the human population.

At its core, neurodiversity promotes the idea that neurological differences, such as autism, ADHD, dyslexia, and intellectual disabilities, should be accepted and respected, rather than stigmatized or pathologized. It acknowledges that individuals with these differences have unique strengths, perspectives, and ways of experiencing the world.

The concept of neurodiversity challenges the dominant neurotypical bias, which assumes that there is a single "normal" or "typical" way of thinking, behaving, and perceiving. This bias often leads to the marginalization and exclusion of neurodivergent individuals, denying them equal opportunities and rights. Neurodiversity advocates argue that we should embrace a more inclusive society that values and accommodates the diverse needs and capabilities of all individuals, regardless of their neurological profile.

One key aspect of understanding neurodiversity is recognizing that it is not limited to a specific set of conditions or diagnoses. While conditions like autism, ADHD, dyslexia, and intellectual disabilities are commonly discussed within the context of neurodiversity, the concept encompasses a much broader range of neurological differences and variations. It acknowledges that neurodivergent traits exist on a continuum, and everyone falls somewhere along that spectrum.

The importance of acceptance and inclusion cannot be overstated when discussing neurodiversity. Acceptance means embracing and valuing neurodivergent individuals for who they are, rather than trying to change or "fix" them. It involves challenging societal norms and attitudes that perpetuate discrimination and prejudice against neurodivergent individuals. Inclusion, on the other hand, focuses on creating environments that are accessible and supportive to neurodivergent individuals, ensuring equal opportunities and the ability to fully participate in all aspects of life.

Celebrating neurodiversity goes beyond mere acceptance and inclusion; it means recognizing and appreciating the unique strengths and contributions that neurodivergent individuals bring to our society. Neurodivergent individuals often exhibit exceptional abilities in areas such as pattern recognition, attention to detail, creativity, problem-solving, and innovation. By embracing neurodiversity, we can tap into these strengths and unlock the full potential of all individuals, leading to greater societal progress and advancement.

Of course, neurodivergent individuals also face challenges and obstacles in navigating a neurotypical world. These challenges can include difficulties with social interaction, sensory sensitivities, communication impairments, executive functioning deficits, and learning differences. It is crucial for society to provide appropriate support, accommodations, and resources to help them overcome these challenges and thrive.

Promoting neurodiversity in society involves a collective effort. It starts with education and raising awareness about neurodiversity and dispelling myths and misconceptions surrounding neurodivergent conditions. It involves creating inclusive educational and work environments that provide appropriate accommodations and support for neurodivergent individuals. It also requires advocating for policy changes that promote equality, accessibility, and inclusion for neurodivergent individuals in all aspects of life.

In debunking neurodiversity myths, it is important to challenge the notion that neurodivergent individuals are broken and need to be "cured." This medical model approach often focuses on deficits and weaknesses, overlooking the strengths and capabilities that neurodivergent individuals possess. Shifting the focus to a neurodiversity paradigm acknowledges that neurodivergent individuals have valuable contributions to make and that their differences should be embraced rather than pathologized.

Identity plays a crucial role in the understanding and acceptance of neurodiversity. Neurodivergent individuals often advocate for embracing their neurodivergent identity and rejecting any attempts to suppress or change who they are. Recognizing neurodivergent identity helps foster a sense of pride, belonging, and empowerment in neurodivergent individuals, promoting their overall well-being and quality of life.

In conclusion, neurodiversity challenges the traditional understanding of neurological differences by celebrating and embracing the natural variation in human brain wiring. It promotes acceptance, inclusion, and recognition of the unique strengths and contributions of neurodivergent individuals. By shifting our focus from deficits to capabilities, we can build a more inclusive and equitable society that values and celebrates neurodiversity in all its forms.

The Neurotypical Bias and Its Impact

The Neurotypical Bias refers to the societal favoritism towards individuals who fit within the typical range of neurocognitive functioning. It is the tendency to value and prioritize neurotypical traits and abilities, while disregarding or devaluing the experiences, strengths, and perspectives of individuals who are neurodivergent.

Neurotypical individuals, often referred to as "typical" or "normal," are those whose neurological development and functioning align with the majority of the population. They usually do not have significant cognitive, sensory, or social differences that would fall under the umbrella of neurodevelopmental conditions.

The impact of the Neurotypical Bias is widespread and affects various domains of life, including education, employment, and social interactions. It perpetuates a system that marginalizes and limits opportunities for neurodivergent individuals, reinforcing societal norms and expectations that may not be applicable or beneficial to their unique strengths and needs.

In education, the Neurotypical Bias can lead to exclusionary practices and inadequate support for neurodivergent students. Traditional teaching methods and curricula often prioritize neurotypical learning styles and communication patterns, leaving neurodivergent learners at a disadvantage. Consequently, these students may face challenges in academic achievement, socialization, and self-esteem.

Similarly, in the workplace, the Neurotypical Bias can present barriers to employment for neurodivergent individuals. Hiring practices and job requirements may inadvertently discriminate against the skills and abilities of neurodivergent individuals, resulting in limited job opportunities and underemployment. This bias not only deprives the workforce of valuable neurodivergent talent but also perpetuates stereotypes and misconceptions about the capabilities of individuals with neurodivergent conditions.

The impact of the Neurotypical Bias also extends to social interactions and relationships. Neurodivergent individuals often experience difficulties in communication and social cues, which can lead to social isolation and exclusion. The emphasis on social norms and expectations may prevent neurodivergent individuals from fully expressing themselves, leading to feelings of being misunderstood or invalidated.

To address the impact of the Neurotypical Bias, it is crucial to promote acceptance and inclusion of neurodivergent individuals. This involves recognizing and valuing the diversity of neurocognitive profiles and challenging the notion that neurotypicality is the norm or ideal. By creating environments that accommodate and celebrate neurodiversity, we can foster a more inclusive society that benefits from the unique perspectives and contributions of neurodivergent individuals.

Educational institutions can adopt inclusive practices that cater to a range of learning styles and abilities. This could involve implementing Universal Design for Learning (UDL) principles, which provide multiple means of representation, engagement, and expression. It also requires professional development opportunities for teachers to enhance their knowledge and understanding of neurodivergent learners' needs.

In the workplace, employers can adopt inclusive hiring practices that focus on skills and potential rather than traditional qualifications. Providing reasonable accommodations, such as flexible work schedules or assistive technologies, can enable neurodivergent individuals to thrive in their chosen professions. Building neurodiversity-friendly workplaces involves fostering a culture of acceptance and providing support for social integration and communication.

Society as a whole can challenge the Neurotypical Bias by promoting awareness and understanding of neurodiversity. This can be achieved through media representation, public campaigns, and advocacy efforts. By amplifying the voices of neurodivergent individuals and sharing their stories, we can challenge stereotypes and misconceptions.

In conclusion, the Neurotypical Bias has far-reaching effects on neurodivergent individuals across various aspects of life. Recognizing and addressing this bias is essential for creating a more inclusive society that values and celebrates neurodiversity. By promoting acceptance, providing appropriate support, and challenging societal norms, we can foster an environment in which neurodivergent individuals can truly thrive.

The Importance of Acceptance and Inclusion

Acceptance and inclusion are fundamental principles of neurodiversity that emphasize the value and worth of all individuals, regardless of their neurodevelopmental differences. In a neurotypical world, where certain traits and behaviors are considered "normal" or "typical," neurodivergent individuals often face marginalization and discrimination. It is therefore crucial to understand and promote the importance of acceptance and inclusion in order to create a more equitable and supportive society.

Defying the Neurotypical Bias

The neurotypical bias refers to the preference for neurotypical individuals, who exhibit typical neurological development, over those who are neurodivergent. This bias is deeply ingrained in many aspects of society, including education, employment, and social interactions. As a result, neurodivergent individuals often experience exclusion, misunderstanding, and stigmatization.

Acceptance and inclusion challenge this bias by recognizing and valuing neurodivergent individuals for who they are, with their unique strengths, perspectives, and contributions. It requires a shift in mindset, replacing the notion of "normality" with an understanding and appreciation of neurological diversity.

Creating a Supportive Environment

Acceptance and inclusion go beyond mere tolerance. They entail creating an environment that supports the needs and experiences of neurodivergent individuals. This means acknowledging and accommodating their unique learning styles, communication preferences, and sensory sensitivities.

In educational settings, acceptance and inclusion involve implementing inclusive practices such as Universal Design for Learning (UDL). UDL recognizes that individuals have diverse ways of processing information and accessing knowledge, and strives to provide multiple means of representation, action and expression, and engagement. By adopting UDL, educators can create a welcoming and inclusive learning environment that benefits all learners, regardless of their neurodivergent status.

Similarly, in the workplace, acceptance and inclusion require employers to provide accommodations and supports that enable neurodivergent individuals to thrive. This can include flexible work arrangements, sensory-friendly workspaces, and clear communication strategies. By embracing the strengths and unique perspectives of neurodivergent employees, organizations can foster innovation, creativity, and productivity.

Addressing Stereotypes and Misconceptions

Acceptance and inclusion also involve challenging stereotypes and misconceptions about neurodivergent individuals. Stereotypes perpetuate harmful biases and can hinder the successful integration of neurodivergent individuals in society. It is essential to debunk these stereotypes and promote accurate understanding and representation.

For example, individuals with Autism Spectrum Disorder (ASD) are often stereotyped as socially awkward or lacking empathy. However, this fails to capture the vast diversity within the autism community and disregards the many strengths and talents individuals with ASD possess, such as attention to detail, pattern recognition, and a unique way of thinking.

By raising awareness, educating the public, and promoting positive narratives about neurodivergent individuals, we can challenge stereotypes and promote acceptance and inclusion. Media representation, literature, and art featuring neurodivergent individuals play a vital role in debunking stereotypes and fostering understanding.

Promoting Empathy and Understanding

Acceptance and inclusion require empathy and understanding from neurotypical individuals, schools, workplaces, and society as a whole. It is crucial to recognize and appreciate the diverse experiences and perspectives of neurodivergent individuals. This entails actively listening to their stories, concerns, and needs, and then taking meaningful action to address them.

For example, neurotypical individuals can educate themselves about the challenges faced by neurodivergent individuals and the support mechanisms that can enhance their well-being. By putting themselves in the shoes of neurodivergent individuals, neurotypical individuals can develop empathy and better advocate for inclusive practices.

Additionally, fostering understanding can be achieved by promoting interpersonal connections and friendships between neurodivergent and neurotypical individuals. Peer support, mentoring programs, and community initiatives can facilitate these connections and break down barriers between different neurological profiles.

Advocacy and Social Change

Acceptance and inclusion are not just individual responsibilities, but also require systemic change and advocacy at the societal level. Advocacy efforts should focus on promoting policies, legislation, and institutional practices that foster acceptance, accommodations, and equal opportunities for neurodivergent individuals.

Neurodivergent individuals themselves play a crucial role in advocacy and bringing about social change. Their voices should be uplifted and amplified to influence policies and attitudes. By sharing their lived experiences, advocating for inclusivity, and participating in decision-making processes, neurodivergent individuals can powerfully contribute to creating a more accepting and inclusive society.

Thinking Beyond Neurodiversity

While acceptance and inclusion are crucial components of the neurodiversity movement, it is important to acknowledge that they should extend beyond the realm of neurodivergent individuals. Inclusivity should encompass all dimensions of diversity, including race, ethnicity, gender, sexuality, and socioeconomic status.

Embracing intersectionality, which recognizes the interconnected nature of different social identities and experiences, allows for a more nuanced understanding of the challenges faced by individuals at the intersections of multiple marginalized identities. By promoting acceptance and inclusion across all dimensions of diversity, we can create a society that celebrates and values the richness and uniqueness of every individual.

Conclusion

Acceptance and inclusion are paramount in promoting a more equitable and compassionate society. By challenging the neurotypical bias, creating supportive environments, addressing stereotypes, promoting empathy and understanding, advocating for change, and embracing intersectionality, we can build a society that celebrates the diversity of human neurology and ensures equal opportunities for all. Embracing neurodiversity enriches our collective experience and fosters innovation, creativity, and social progress. It is only by dismantling barriers and embracing the strengths and unique perspectives of neurodivergent individuals that we can truly celebrate differences in a neurotypical world.

Why Celebrate Neurodiversity?

Neurodiversity is a concept that recognizes and celebrates the natural variation in neurological traits and abilities among individuals. It encourages society to embrace and accept this diversity rather than viewing it as a deviation from the norm. In this section, we will explore the reasons why celebrating neurodiversity is important and beneficial for individuals and society as a whole.

Promoting Inclusion

One of the key reasons to celebrate neurodiversity is to promote inclusion. By acknowledging and accepting individuals with neurodivergent traits, we create an environment where everyone feels valued and included. Embracing neurodiversity means moving away from a mindset that favors a neurotypical ideal and recognizing the unique contributions that neurodivergent individuals can bring. Just like any other form of diversity, such as race or gender, neurodiversity adds richness and depth to our communities and workplaces.

Unlocking Hidden Talents

Neurodiversity is not just about recognizing differences; it is also about acknowledging the strengths and talents that come with those differences. Many neurodivergent individuals possess exceptional abilities in areas such as pattern recognition, attention to detail, problem-solving, creativity, and innovative thinking. By celebrating neurodiversity, we provide a platform for these talents to shine and contribute to various fields of study, research, and innovation. By harnessing and supporting these hidden talents, we can foster a more inclusive and productive society.

Challenging Stereotypes and Stigma

Celebrating neurodiversity helps challenge stereotypes and reduce the stigma associated with neurodivergent conditions. Stereotypes often perpetuate misconceptions and discrimination, leading to barriers in education, employment, and social interactions. By highlighting the diverse strengths and capabilities of neurodivergent individuals, we break down these stereotypes and promote a more accurate and positive understanding of neurodiversity. This shift in perspective can lead to increased acceptance and support for individuals with neurodivergent traits.

Fostering Empathy and Understanding

Neurodiversity celebration fosters empathy and understanding among individuals. It allows us to recognize that each person's experience of the world is unique and valid, regardless of their neurological makeup. By embracing neurodiversity, we develop a broader perspective and a greater capacity to empathize with others, leading to a more compassionate and inclusive society. This empathy extends beyond neurodivergent individuals to all individuals who experience differences in any aspect of their lives.

Encouraging Neurodivergent Identity Development

Celebrating neurodiversity plays a crucial role in fostering a positive sense of identity in neurodivergent individuals. It helps them understand that their neurodivergent traits are not defects or weaknesses but valuable aspects of their identity. By providing a platform to celebrate their uniqueness, we empower neurodivergent individuals to embrace their strengths, build self-esteem, and advocate for their needs. This self-acceptance and pride in neurodivergent identity create a positive ripple effect, leading to increased confidence and overall well-being.

Promoting Innovation and Creativity

Neurodiversity celebration fuels innovation and creativity. Neurodivergent individuals often possess a different way of thinking, processing information, and approaching problems. These unique cognitive styles can lead to fresh perspectives and out-of-the-box solutions. By including neurodivergent individuals in diverse teams and embracing their neurodivergent perspectives, we create an environment that nurtures innovation and fosters creative problem-solving. Celebrating neurodiversity has the potential to unlock new ideas, discoveries, and breakthroughs in various domains.

In conclusion, celebrating neurodiversity is essential for promoting inclusion, unlocking hidden talents, challenging stereotypes, fostering empathy, encouraging identity development, and fueling innovation. By embracing and celebrating neurodiversity, we create a society that values and respects the unique contributions of all individuals, regardless of their neurological differences. Let us celebrate neurodiversity and build a more inclusive and prosperous future for everyone.

Challenges Faced by Neurodivergent Individuals

Neurodivergent individuals face a range of challenges in their daily lives, stemming from the differences in their neurodevelopmental conditions. These challenges can manifest in various aspects of their lives, including social interactions, communication, education, and employment. Understanding and addressing these challenges is crucial for creating a more inclusive and supportive society. In this section, we will explore some of the key challenges faced by neurodivergent individuals and discuss strategies to address them.

1. Sensory Sensitivities: Many neurodivergent individuals experience heightened sensory sensitivities. They may be hypersensitive or hyposensitive to various sensory inputs, including sound, light, touch, taste, and smell. Sensory overload can lead to anxiety, discomfort, and difficulty focusing. To address this challenge, it is important to create sensory-friendly environments that minimize sensory stimuli and provide support for self-regulation. This can include the use of noise-cancelling headphones, dimmed lighting, designated quiet spaces, and sensory tools like fidget toys.

2. Social Interactions and Communication: Neurodivergent individuals often face challenges in social interactions and communication. They may have difficulty understanding and interpreting nonverbal cues, such as facial expressions and body language. This can lead to social anxiety, isolation, and difficulties in forming and maintaining relationships. To address this challenge, education and awareness about neurodiversity are essential. Providing clear and explicit communication strategies, such as visual supports, social stories, and social skills training, can help neurodivergent individuals navigate social interactions more effectively.

3. Executive Functioning: Many neurodivergent individuals struggle with executive functioning skills, which involve processes such as planning, organizing, time management, and decision-making. Difficulties in these areas can affect academic performance, work productivity, and daily life skills. To support individuals with executive functioning challenges, it is important to provide structured routines, visual schedules, task breakdowns, and explicit instruction in organization and time management strategies. Additionally, assistive technologies, such as reminder apps and organization tools, can be beneficial.

4. Emotional Regulation: Neurodivergent individuals may face challenges in regulating their emotions, leading to difficulties in managing stress, anxiety, and frustration. They may have intense emotional reactions to certain situations or struggle with identifying and expressing their emotions. Teaching emotional regulation strategies, such as deep breathing exercises, mindfulness techniques, and cognitive-behavioral approaches, can be helpful in managing emotions effectively. Creating safe spaces for emotional expression and providing support networks are also important.

5. Academic and Work-related Challenges: Neurodivergent individuals may encounter specific challenges in academic and work settings. Difficulties in reading, writing, and understanding verbal instructions can impact academic performance. Workplace challenges may include difficulties in task completion, time management, and sensory overload. Accommodations, such as extended time for

exams, assistive technology tools, flexible work arrangements, and workplace accommodations, can help mitigate these challenges and support neurodivergent individuals in their academic and professional pursuits.

6. Mental Health and Well-being: Neurodivergent individuals are at a higher risk of experiencing mental health issues, such as anxiety, depression, and emotional dysregulation. These challenges can be attributed to a range of factors, including societal stigma, difficulties in navigating social relationships, and coping with the demands of daily life. Providing accessible mental health services and supports, promoting self-acceptance and self-care practices, and fostering a supportive community are vital for promoting the mental health and well-being of neurodivergent individuals.

It is important to note that the challenges faced by neurodivergent individuals can vary widely depending on their specific condition and individual experiences. A holistic approach that considers each individual's unique strengths and needs is crucial for effective support and inclusion. By acknowledging and addressing these challenges, we can create a more inclusive society that celebrates and embraces the diversity of neurodivergent individuals.

Resources: 1. National Autistic Society. (n.d.). Sensory sensitivities. Retrieved from `https://www.autism.org.uk/about/behaviour/sensory-world.aspx` 2. Autism Society. (n.d.). Social skills. Retrieved from `https://www.autism-society.org/living-with-autism/how-the-autism-society-can-help/social-skills/` 3. Understood. (2021). What is executive functioning? Retrieved from `https://www.understood.org/en/learning-thinking-differences/child-learning-disabilities/executive-functioning-issues/understanding-executive-functioning-issues` 4. Child Mind Institute. (n.d.). Emotional regulation. Retrieved from `https://childmind.org/guide/guide-to-telehealth/understanding-mental-health/emotional-regulation/` 5. Job Accommodation Network. (n.d.). Accommodations for specific neurodiverse conditions. Retrieved from `https://askjan.org/topics/neurodiverse.cfm` 6. National Institute of Mental Health. (n.d.). Mental health information. Retrieved from `https://www.nimh.nih.gov/health/topics/children-and-mental-health/index.shtml`

Promoting Neurodiversity in Society

Promoting neurodiversity in society is crucial for creating an inclusive and accepting environment for individuals with neurodevelopmental conditions. It involves challenging biases, dispelling myths, and advocating for the rights and well-being of neurodivergent individuals. By promoting neurodiversity, we can foster a society that celebrates the unique strengths and perspectives of everyone, regardless of their neurological differences.

Challenging Biases and Stereotypes

One of the first steps in promoting neurodiversity is challenging biases and stereotypes associated with neurodevelopmental conditions. Neurodivergent individuals often face misconceptions and negative assumptions about their abilities. As a society, we need to recognize and address these biases, replacing them with accurate and positive perceptions.

Education plays a crucial role in challenging biases. Schools, workplaces, and communities should provide accurate information about various neurodevelopmental conditions to promote understanding and empathy. By debunking common myths surrounding neurodivergent individuals, we can create a more accepting society.

Advocacy and Awareness Raising

Another key aspect of promoting neurodiversity is advocacy and raising awareness. Advocacy involves speaking up for the rights and needs of neurodivergent individuals, striving for equal opportunities, and dismantling barriers to inclusion. It also involves promoting the understanding that neurodivergent individuals have unique strengths and talents that can contribute to society.

Awareness campaigns and initiatives can help bring attention to the challenges faced by neurodivergent individuals and the importance of creating inclusive spaces. These campaigns can be targeted towards schools, workplaces, healthcare settings, and the general public. They can highlight the benefits of neurodiversity and provide resources for support and understanding.

Inclusive Policies and Practices

Promoting neurodiversity requires implementing inclusive policies and practices in various settings. In schools, this may involve adopting individualized education plans (IEPs) to accommodate diverse learning needs and providing necessary supports and accommodations. It can also involve incorporating universal design for learning (UDL) principles to create inclusive classroom environments that cater to different learning styles.

In workplaces, inclusive policies can include flexible work arrangements, reasonable accommodations, and diversity training for employees. Employers can actively recruit and support neurodivergent individuals, recognizing their unique skills and contributions. By fostering an inclusive work environment, organizations can benefit from the diverse perspectives and problem-solving abilities of neurodivergent employees.

Collaboration and Partnerships

Promoting neurodiversity requires collaboration and partnerships between various stakeholders, including individuals, families, educators, healthcare professionals, and community organizations. By working together, we can create a supportive network that provides resources, guidance, and opportunities for neurodivergent individuals.

Collaboration can involve forming support groups, organizing educational workshops, and creating mentorship programs. It can also involve partnering with advocacy organizations and community groups to amplify the voices of neurodivergent individuals and advocate for systemic change.

Creating Inclusive Environments

Creating inclusive environments is a fundamental aspect of promoting neurodiversity in society. In schools, this may involve providing sensory-friendly spaces, promoting positive social interactions, and fostering a culture of acceptance and understanding. Workplaces can create inclusive environments by implementing sensory-friendly design elements, providing clear communication channels, and promoting a culture of diversity and inclusion.

Additionally, creating inclusive environments means valuing and respecting the individual identities and experiences of neurodivergent individuals. It requires embracing diverse communication styles and providing opportunities for self-expression and self-advocacy.

Thinking Beyond Labels

Promoting neurodiversity in society also involves shifting our mindset from focusing solely on diagnostic labels to understanding and embracing the unique strengths and challenges of each individual. It means recognizing that neurodivergent individuals have a wide range of abilities, skills, and talents that should be valued and nurtured.

By thinking beyond labels, we can break down barriers and create opportunities for neurodivergent individuals to thrive. This can involve providing diverse educational and employment pathways, celebrating neurodivergent achievements, and promoting inclusive practices that honor the contributions of all individuals, regardless of their neurological profile.

Embracing Neurodiversity as a Social Value

Ultimately, promoting neurodiversity in society requires embracing it as a social value. It entails recognizing that neurodiversity is an inherent part of human diversity, and that our society is enriched by the different ways in which individuals experience and engage with the world.

By promoting neurodiversity, we can foster a society that acknowledges and appreciates the unique contributions of neurodivergent individuals. It enables us to move beyond the neurotypical bias and build a more inclusive, accepting, and flourishing society for all.

Debunking Neurodiversity Myths

Neurodiversity is a concept that challenges our traditional understanding of neurological differences and seeks to celebrate the diversity of human brains. However, there are several myths and misconceptions surrounding neurodiversity that need to be debunked. In this section, we will explore and address some of these myths, providing evidence and insights to help foster a better understanding and appreciation of neurodiversity.

Myth 1: Neurodivergent individuals are just seeking attention

One common myth about neurodiversity is that individuals who are neurodivergent are deliberately seeking attention by emphasizing their differences. This misconception stems from a lack of understanding about the challenges and experiences faced by neurodivergent individuals. In reality, neurodivergent individuals often struggle with social interactions, sensory sensitivities, and other aspects of daily life that neurotypical individuals may take for granted. Their behaviors and needs are not motivated by a desire for attention, but rather a genuine expression of their neurological differences.

Myth 2: Neurodivergent individuals are lazy or unmotivated

Another myth surrounding neurodiversity is that neurodivergent individuals are lazy or lack motivation. This misconception fails to recognize the unique strengths and challenges that come with different neurological profiles. Neurodivergent individuals may have difficulties with executive functions, such as organization and time management, which can affect their ability to complete tasks efficiently. However, it is important to understand that this does not stem from laziness or a lack of motivation. Rather, it reflects differences in cognitive processing and the need for tailored support and accommodations.

Myth 3: Neurodiversity is just an excuse for bad behavior

Neurodiversity is often misunderstood as an excuse for "bad" or disruptive behavior. This myth fails to recognize that certain behaviors exhibited by neurodivergent individuals may be a result of their neurological differences rather than intentional misbehavior. For example, individuals with autism may have difficulty with sensory processing, leading to sensory overload and meltdowns. It is essential to understand these behaviors within the context of neurodiversity and provide appropriate support and understanding rather than labeling them as "bad" or seeking punishment.

Myth 4: Neurodivergent individuals cannot succeed academically or professionally

There is a pervasive myth that neurodivergent individuals cannot excel academically or professionally. This misconception ignores the unique strengths often associated with different neurodivergent conditions. For example, individuals with ADHD may excel in tasks that require flexibility and creative problem-solving. Similarly, individuals on the autism spectrum may have exceptional attention to detail and focus, making them well-suited for certain professions. It is crucial to understand that neurodivergent individuals may have different ways of learning and working, and with the right support and accommodations, they can achieve great success.

Myth 5: Neurodiversity is a fad or trend

Some people view the concept of neurodiversity as a passing fad or trend. However, neurodiversity is rooted in the recognition of the inherent value of diverse brains and the promotion of inclusivity and acceptance. It is not a temporary trend but a vital perspective that challenges the dominant neurotypical bias in our society. By embracing neurodiversity, we can create a more inclusive and equitable world for all individuals, irrespective of their neurological differences.

In debunking these myths, it is important to recognize the diversity within the neurodivergent community itself. Each individual's experience and strengths may vary depending on their specific neurodivergent condition. It is essential to approach neurodiversity with an open mind, seek to understand the unique perspectives and needs of neurodivergent individuals, and promote acceptance and inclusion in all aspects of life.

Key Takeaways:

- Neurodivergent individuals are not seeking attention but rather expressing their neurological differences.

- Laziness or lack of motivation should not be attributed to neurodivergent individuals, as their challenges are rooted in differences in cognitive processing.

- Neurodiversity is not an excuse for bad behavior but should be understood within the context of neurological differences.

- Neurodivergent individuals can excel academically and professionally with the right support and accommodations.

- Neurodiversity is not a passing trend but a vital perspective that challenges the neurotypical bias in society.

Exercises

1. Identify a neurodivergent myth or misconception that you have encountered in your own life or community. Explain why this myth is inaccurate and provide evidence or personal anecdotes to debunk it.

2. Conduct research on successful neurodivergent individuals in various fields, such as science, arts, or entrepreneurship. Write a short report highlighting their achievements and the strengths associated with their neurodivergent condition.

3. Imagine you are an educator or employer wanting to create a neurodiversity-friendly environment. Develop a list of strategies and accommodations you would implement to support the success and well-being of neurodivergent individuals.

Resources

- "Neurotribes: The Legacy of Autism and the Future of Neurodiversity" by Steve Silberman. - "Neurodiversity: Discovering the Extraordinary Gifts of Autism, ADHD, Dyslexia, and Other Brain Differences" by Thomas Armstrong. - "Loud Hands: Autistic People, Speaking" edited by Julia Bascom. - The Autistic Self Advocacy Network (`https://autisticadvocacy.org/`) - The National Center for Learning Disabilities (`https://www.ncld.org/`) - The ADHD Foundation (`https://www.adhdfoundation.org.uk/`)

Neurodiversity Paradigm vs Medical Model Approach

In understanding neurodiversity, it is crucial to explore the contrasting philosophies of the neurodiversity paradigm and the medical model approach. These two perspectives shape our understanding of neurodivergent individuals and have significant implications for how we support and include them in society.

The medical model approach traditionally views neurodivergent conditions as disorders or disabilities that require intervention, treatment, and a focus on "normalizing" individuals to fit predetermined standards. This approach emphasizes identifying deficits and impairments and seeks to find ways to mitigate or fix them. In this model, the emphasis is placed on the individual's deficits rather than their strengths, and their differences are pathologized.

On the other hand, the neurodiversity paradigm challenges this medical model by asserting that neurodivergent conditions are natural variations of the human brain rather than solely deficits or disorders. It celebrates the diversity of neurological traits and rejects the idea that there is a single "normal" way of being. The neurodiversity paradigm recognizes that neurodivergent individuals have unique strengths and abilities that should be valued and accommodated rather than suppressed or "fixed."

In the neurodiversity paradigm, the focus shifts from trying to change individuals into fitting societal norms to promoting acceptance, understanding, and inclusion. It advocates for accommodating and supporting neurodivergent individuals in all aspects of life, including education, employment, and social settings. The goal is to create a society that values and embraces different ways of thinking, processing information, and experiencing the world.

One key aspect of the neurodiversity paradigm is the concept of neurodivergent identity. It acknowledges that neurodivergent individuals have a distinct identity shaped by their neurological differences. This identity is not seen as a deficiency or a problem but as an essential part of who they are. Neurodivergent individuals are encouraged to embrace and celebrate their unique experiences, perspectives, and abilities.

The neurodiversity paradigm challenges the negative stereotypes and stigma associated with neurodivergent conditions. It emphasizes the importance of destigmatizing neurodiversity and creating a more inclusive and accepting society. By shifting the narrative from deficit-based thinking to one that recognizes and values diversity, the neurodiversity paradigm offers a more empowering and respectful approach to supporting neurodivergent individuals.

While the medical model still has its place in providing necessary healthcare interventions and support, the neurodiversity paradigm enables a broader understanding and acceptance of neurodivergent individuals. It recognizes that their differences contribute to the rich tapestry of human diversity and that an inclusive society is one in which every individual is valued and respected for who they are.

The shift from the medical model approach to the neurodiversity paradigm requires changes at various levels, including policy, education, and societal attitudes. It necessitates the creation of supportive environments, the development of inclusive educational programs, and the provision of appropriate accommodations in the workplace. Breaking down barriers and promoting acceptance and inclusion are essential for the full participation and well-being of neurodivergent individuals.

To truly embrace the neurodiversity paradigm, it is crucial for society to challenge its biases, educate themselves about different neurodivergent conditions, and actively work towards creating a more inclusive and accepting world. By doing so, we can celebrate the unique strengths and contributions of neurodivergent individuals and create a society that values and supports the neurodiversity within it.

Real-world Example:

One real-world example of the neurodiversity paradigm in action is the movement to improve neurodiversity inclusion in the workplace. Instead of viewing neurodivergent individuals as liabilities or solely focusing on their challenges, some forward-thinking companies recognize the valuable skills and different perspectives they bring to the table.

For instance, software companies like SAP and Microsoft have initiated neurodiversity hiring programs, actively recruiting and employing individuals on the autism spectrum. These companies are leveraging the strengths of neurodivergent individuals by offering job roles that align with their unique skills, such as data analysis, quality assurance, and software testing.

By adopting the neurodiversity paradigm, these companies are not only providing employment opportunities to neurodivergent individuals but are also redefining workplace norms and fostering a culture of diversity and inclusion. The accomplishments and successes of these neurodivergent employees challenge the traditional notion of what constitutes a valuable employee and promote a more inclusive work environment.

Key Takeaways:

- The medical model approach views neurodivergent conditions as disorders or disabilities, while the neurodiversity paradigm celebrates them as natural variations of the human brain.
- The neurodiversity paradigm emphasizes acceptance, understanding, and inclusion of neurodivergent individuals, focusing on their strengths and unique abilities.
- Neurodivergent identity is a core concept within the neurodiversity paradigm, recognizing and celebrating the distinct experiences and perspectives of neurodivergent individuals.
- Shifting from the medical model approach to the neurodiversity paradigm requires changes at various levels, including policy, education, and societal attitudes.
- Real-world examples, such as neurodiversity hiring programs in companies, highlight the practical application and benefits of embracing the neurodiversity paradigm.

Further Resources:

1. Armstrong, T. (2010). *Neurodiversity in the classroom: Strength-based strategies to help students with special needs succeed in school and life.* ASCD.
2. Silberman, S. (2015). *NeuroTribes: The Legacy of Autism and the Future of Neurodiversity.* Avery.
3. The National Symposium on Neurodiversity at Syracuse University (neurodiversity.syr.edu)

In conclusion, understanding the contrasting perspectives of the neurodiversity paradigm and the medical model approach is critical for fostering a more inclusive and accepting society. By embracing the neurodiversity paradigm and recognizing the strengths and value of neurodivergent individuals, we can create a world where diversity is truly celebrated and all individuals are given the opportunity to thrive.

The Role of Identity in Neurodiversity

Identity plays a crucial role in the understanding and celebration of neurodiversity. It encompasses how individuals perceive themselves and how they are perceived by others within the context of their neurodivergent traits. Neurodiverse individuals, such as those with autism, ADHD, dyslexia, and intellectual disabilities, often face stigma and societal expectations that can impact their sense of self. Recognizing and embracing one's neurodivergent identity is essential for fostering acceptance, inclusion, and empowerment.

Neurodivergent individuals often encounter challenges related to their identity, including societal misconceptions, stereotyping, and discrimination. These negative experiences can lead to feelings of exclusion, low self-esteem, and a sense of not fitting into the neurotypical world. However, embracing one's neurodivergent identity can counteract these challenges by promoting self-acceptance and valuing the unique strengths and perspectives that neurodivergent individuals bring to society.

One key aspect of neurodivergent identity is the concept of neurodivergent pride. This notion emphasizes the importance of valuing and celebrating neurodivergent traits rather than viewing them as deficits or problems to be fixed. Neurodivergent pride encourages individuals to embrace their differences and reject the notion that there is a 'normal' or 'ideal' neurotypical way of being. It empowers neurodivergent individuals to advocate for themselves, challenge societal norms, and redefine what it means to be successful in a neurotypical world.

A critical component of developing a positive neurodivergent identity is self-awareness. Understanding one's own strengths, abilities, challenges, and interests is essential for building a strong sense of self. This self-awareness helps individuals recognize and appreciate the unique contribution they can make to society. It encourages them to assert their needs, express themselves authentically, and pursue paths that align with their passions and abilities.

Identity also extends beyond the individual and encompasses the concept of community. Neurodivergent individuals can find support and connection within neurodivergent communities, where they can share their experiences, concerns, and triumphs with others who understand and relate to their neurodivergent identity. These communities provide a safe space for individuals to explore and affirm their neurodivergent identities, foster a sense of belonging, and amplify their collective voices for advocacy and change.

It is important to note that neurodivergent identity is multidimensional and intersects with other aspects of a person's identity, such as race, gender, sexuality, and culture. Intersectionality acknowledges that different forms of identity interact and influence an individual's experiences and opportunities. Understanding and addressing these intersections fosters a more inclusive and nuanced understanding of neurodiversity.

In order to support the role of identity in neurodiversity, it is crucial to promote a culture of acceptance and inclusion. This involves challenging ableism, advocating for neurodivergent individuals' rights, and creating environments that recognize and accommodate the diverse needs of all individuals. Schools, workplaces, and communities can implement policies and practices that respect and value neurodivergent identities, promoting a sense of belonging and enabling neurodivergent individuals to thrive.

In conclusion, the role of identity in neurodiversity is paramount. Recognizing and embracing one's neurodivergent identity allows individuals to develop self-acceptance, advocate for their needs, and redefine success on their own terms. It fosters a sense of pride in neurodivergent traits, enables individuals to connect with supportive communities, and promotes a more inclusive and accepting society. By valuing neurodivergent identities, we can celebrate the diversity of the human experience and work towards a more equitable and inclusive world for all.

Neurodevelopmental Conditions

Autism Spectrum Disorder (ASD)

Autism Spectrum Disorder (ASD) is a neurodevelopmental condition that affects individuals in various ways, resulting in differences in social communication, sensory processing, and behavioral patterns. It is characterized by a range of challenges and strengths, creating a unique profile for each individual with ASD. In this section, we will explore the definition and diagnosis of ASD, understanding the characteristics and challenges faced by individuals with ASD, and the importance of supporting them in educational and work settings.

Definition and Diagnosis of ASD

ASD is a complex neurodevelopmental condition that affects how individuals perceive and interact with the world around them. It is typically diagnosed in early childhood, although it can be identified in adults as well. The diagnostic criteria for ASD are outlined in the Diagnostic and Statistical Manual of Mental Disorders (DSM-5), a widely used classification system in psychiatry.

According to the DSM-5, ASD is characterized by persistent deficits in social communication and social interaction across multiple contexts. This may manifest as challenges in initiating and maintaining conversations, difficulties understanding nonverbal cues, and limited or repetitive patterns of behavior, interests, or activities. Individuals with ASD may also display sensory sensitivities, whereby they may be over- or under-responsive to certain sensory stimuli, such as light, sound, or touch.

The diagnosis of ASD involves a comprehensive assessment conducted by a multidisciplinary team, including psychologists, psychiatrists, speech-language pathologists, and occupational therapists. The assessment typically includes a thorough evaluation of the individual's developmental history, observation of their social and communication skills, and a variety of standardized assessments to measure their abilities and challenges in different areas.

Understanding the Characteristics and Challenges of ASD

ASD is characterized by a wide range of characteristics and challenges that can vary significantly from person to person. While it is essential to recognize and respect the individual differences within the autism community, there are some common features that are often associated with ASD.

1. Social Communication Challenges: Individuals with ASD may have difficulties with social communication skills, such as engaging in reciprocal conversations, understanding and using nonverbal cues (e.g., eye contact, facial expressions), and interpreting the perspective of others. These challenges can impact their ability to form and maintain friendships and navigate social interactions effectively.

2. Restricted and Repetitive Behaviors and Interests: Individuals with ASD may exhibit restricted and repetitive patterns of behavior, interests, or activities. They may have intense and narrow interests in specific topics or objects, engage in repetitive and stereotyped behaviors (e.g., hand-flapping, rocking), or display rigid adherence to routines or rituals. These behaviors and interests can provide a sense of predictability and security for individuals with ASD.

3. Sensory Sensitivities: Many individuals with ASD have sensory sensitivities, meaning they may be hypersensitive or hyposensitive to certain sensory stimuli. They may be overwhelmed by loud noises, bright lights, or certain textures, or seek out intense sensory input to regulate their arousal levels. Sensory sensitivities can significantly impact their daily functioning and may require strategies and accommodations to manage.

4. Executive Functioning Difficulties: Executive functions refer to a set of cognitive processes responsible for planning, organizing, problem-solving, and self-regulation. Individuals with ASD may experience challenges in these areas, including difficulties with organization, time management, flexibility, and impulse

control. These difficulties can impact their academic performance, daily routines, and ability to navigate novel situations.

It is important to note that individuals with ASD also possess unique strengths and abilities. Many individuals with ASD have exceptional memory skills, attention to detail, and a unique way of thinking that can contribute to creative problem-solving and innovation.

Supporting Individuals with ASD in Educational and Work Settings

Creating inclusive and supportive environments is crucial for individuals with ASD to thrive in educational and work settings. By understanding their strengths, challenges, and individual needs, we can provide the necessary support and accommodations to ensure their success. Here are some strategies and considerations for supporting individuals with ASD:

1. Individualized Education Plans (IEPs): In educational settings, individualized education plans are often developed to address the specific needs of students with ASD. These plans outline goals, accommodations, and support strategies to facilitate learning and social integration. Collaboration among teachers, parents, and other specialists is essential in creating and implementing effective IEPs.

2. Universal Design for Learning (UDL): UDL is an educational framework that advocates for providing multiple means of representation, expression, and engagement to meet the diverse learning needs of all students. It promotes flexibility in teaching methods, materials, and assessments, making the learning experience accessible and engaging for individuals with ASD.

3. Social Skills Training: Social skills training programs can be beneficial for individuals with ASD, as they provide explicit instruction and practice opportunities to develop social communication and interaction skills. These programs focus on areas such as initiating and maintaining conversations, understanding social cues, and problem-solving in social situations.

4. Sensory Supports and Accommodations: Sensory sensitivities are common among individuals with ASD. Providing sensory supports and accommodations, such as designated quiet spaces, noise-canceling headphones, or visual schedules, can help individuals regulate their sensory experiences and reduce anxiety or overload.

5. Workplace Accessibility and Accommodations: In work settings, it is essential to create an inclusive and accessible environment for individuals with ASD. This includes providing clear expectations and instructions, offering

flexibility in work arrangements or schedules, and implementing sensory-friendly measures, such as minimizing sensory distractions or providing quiet workspaces.

6. Peer Support and Mentoring: Peer support programs and mentoring initiatives can be beneficial for individuals with ASD, as they provide opportunities for social connection, skill-building, and guidance. Pairing individuals with ASD with neurotypical peers or mentors can foster social inclusion, enhance self-esteem, and promote positive outcomes in educational and work settings.

By embracing the strengths and supporting the unique needs of individuals with ASD, we can create inclusive environments that celebrate neurodiversity and enable individuals with ASD to reach their full potential.

Real-World Example: One of the well-known technological innovators with ASD is Temple Grandin, an animal scientist and professor at Colorado State University. Despite her challenges with social communication and sensory sensitivities, Grandin has made significant contributions to the field of animal behavior and welfare. Her unique perspective and ability to think in pictures have led to revolutionary innovations in animal handling and designing livestock facilities that prioritize animal comfort and reduce stress. Grandin's story showcases the importance of recognizing and harnessing the strengths of individuals with ASD and providing the necessary supports and accommodations for them to excel in their chosen field.

Resources: - "Thinking in Pictures: My Life with Autism" by Temple Grandin - Autism Society: www.autism-society.org - Autism Speaks: www.autismspeaks.org - National Autistic Society: www.autism.org.uk

Summary

In this section, we explored Autism Spectrum Disorder (ASD) and its characteristics. We discussed the definition and diagnosis of ASD, understanding the challenges faced by individuals with ASD, and the importance of supporting them in educational and work settings. We highlighted the importance of individualized support, such as individualized education plans and accommodations, and the value of creating inclusive and accessible environments to promote the success of individuals with ASD. By recognizing and embracing the strengths of individuals with ASD and providing the necessary support, we can celebrate neurodiversity and create a more inclusive and accepting society.

Characteristics and Diagnosis of ASD

Autism Spectrum Disorder (ASD) is a neurodevelopmental condition characterized by persistent challenges in social communication, restricted and repetitive patterns of behavior, interests, or activities. The diagnosis of ASD involves a comprehensive assessment of an individual's behavioral, social, and communication patterns.

Social Communication Challenges

One of the key characteristics of ASD is difficulty in social communication. Individuals with ASD may have challenges in understanding and using verbal and nonverbal cues in conversation. They may struggle with initiating and maintaining conversations, taking turns, and understanding the perspective of others. These difficulties may lead to a preference for solitary activities and limited social interactions.

Restricted and Repetitive Behaviors and Interests

Individuals with ASD often exhibit a range of restricted and repetitive behaviors, which can manifest in various ways. Some common examples include repetitive movements or behaviors (e.g., hand flapping, rocking), insistence on sameness or routines, intense preoccupation with specific interests, and sensory sensitivities (e.g., being overly sensitive or under-responsive to sensory stimuli). These behaviors and interests may be inflexible and interfere with daily functioning.

Sensory Sensitivities

Many individuals with ASD have atypical sensory processing, experiencing hypersensitivity or hyposensitivity to different sensory inputs such as sight, sound, touch, taste, and smell. For example, they may be overwhelmed by certain sounds or textures that others find tolerable, or they may seek out intense sensory experiences. These sensory sensitivities can affect their daily lives and may contribute to behavioral challenges.

Diagnosis of ASD

The diagnosis of ASD involves a comprehensive evaluation conducted by multidisciplinary professionals, such as psychologists, psychiatrists, speech-language pathologists, and pediatricians. The assessment typically includes:

- Detailed developmental history: Gathering information about early childhood milestones, social interactions, communication skills, and any behavioral concerns.

- Direct observation: Assessing the individual's behaviors and interactions in different settings, such as home, school, or clinical environments.

- Standardized assessments: Using validated assessment tools, such as the Autism Diagnostic Observation Schedule (ADOS) and the Autism Diagnostic Interview - Revised (ADI-R), to gather information about social communication, behavior, and developmental history.

- Collaboration with caregivers and teachers: Gathering information from parents, caregivers, and educators about the individual's behavior and social functioning across different settings.

The assessment aims to identify the presence of core symptoms of ASD, their severity, and their impact on daily functioning. It also considers the possibility of co-occurring conditions, such as intellectual disabilities, language impairments, or mental health issues that may require additional evaluation and support.

Early Identification and Screening

Early identification and intervention can significantly improve outcomes for individuals with ASD. Several screening tools are available to identify potential signs of ASD in young children. These tools, such as the Modified Checklist for Autism in Toddlers (M-CHAT) and the Social Communication Questionnaire (SCQ), can be used to identify red flags and warrant further assessment. It's important to note that screening tools do not provide a diagnosis but serve as a way to identify children who may need further evaluation.

Differential Diagnosis

ASD can sometimes present with overlapping characteristics with other neurodevelopmental conditions, such as attention-deficit/hyperactivity disorder (ADHD), intellectual disabilities, specific learning disorders, or language impairments. A thorough assessment is needed to differentiate ASD from other conditions and determine the most appropriate interventions and supports.

Strengths and Challenges

While individuals with ASD face various challenges, it is important to recognize and celebrate their unique strengths. Many individuals with ASD have exceptional abilities in areas such as pattern recognition, attention to detail, or creative thinking. By focusing on their strengths, supportive environments can be created to nurture their talents and enhance their overall well-being.

Resources and Support

Once diagnosed with ASD, individuals and their families can access a range of resources and support systems. These may include early intervention programs, specialized educational services, therapeutic interventions (e.g., speech therapy, occupational therapy), and support from community organizations and advocacy groups. It is essential to provide holistic support that addresses the unique needs of individuals with ASD and promotes their inclusion and participation in society.

In conclusion, Autism Spectrum Disorder (ASD) is characterized by challenges in social communication, restricted and repetitive behaviors, and sensory sensitivities. The diagnosis involves a comprehensive assessment of an individual's developmental history, direct observation, standardized assessments, and collaboration with caregivers and teachers. Early identification and intervention are crucial for improving outcomes, and individuals with ASD have unique strengths that should be recognized and nurtured. Access to resources and support systems can help individuals with ASD thrive and participate fully in society.

Strengths and Challenges of Individuals with ASD

Individuals with Autism Spectrum Disorder (ASD) possess a unique set of strengths and face specific challenges that shape their experiences and interactions with the world. Understanding these strengths and challenges is crucial for fostering acceptance, inclusion, and support for individuals with ASD. In this section, we will explore the different aspects that contribute to the strengths and challenges of individuals with ASD.

Strengths

1. **Attention to Detail:** One of the notable strengths of individuals with ASD is their exceptional attention to detail. They often have an excellent ability to focus on specific tasks, noticing even the smallest details that others may overlook. This

attention to detail can be an asset in various domains, such as scientific research, programming, and art, where precision and accuracy are essential.

2. **Visual Thinking:** Individuals with ASD often have a unique way of thinking and processing information, relying heavily on visual cues. They are adept at visualizing concepts, patterns, and relationships, which can contribute to their creativity and problem-solving abilities. This visual thinking style can be advantageous in fields such as graphic design, architecture, and mathematics.

3. **Memory Skills:** Many individuals with ASD exhibit remarkable memory skills, particularly in areas of specific interest. They can retain vast amounts of information and recall it with remarkable accuracy. This strength can be utilized in professions where extensive knowledge and recall are critical, such as historians, scientists, and researchers.

4. **Analytical Thinking:** Individuals with ASD often demonstrate strong analytical skills, enabling them to break down complex problems into manageable components. Their ability to think logically and systematically can be an asset in fields such as computer programming, engineering, and mathematics.

5. **Hyperfocus:** Hyperfocus refers to the ability of individuals with ASD to become fully engrossed in a task or topic of interest for an extended period. This intense level of concentration allows them to immerse themselves in activities they are passionate about, leading to remarkable achievements and contributions to their respective fields.

Challenges

1. **Social Communication:** One of the primary challenges faced by individuals with ASD is difficulties in social communication and interaction. They may struggle with understanding and using nonverbal cues, such as facial expressions, body language, and tone of voice. This challenge can impact their ability to build and maintain meaningful relationships and navigate social situations effectively.

2. **Sensory Sensitivities:** Many individuals with ASD experience sensitivities to sensory stimuli, such as sounds, lights, textures, or smells. These sensitivities can cause discomfort, overwhelm, or even physical pain, making it challenging to function in environments that may be considered normal to others.

3. **Executive Functioning:** Executive functioning encompasses skills such as planning, organization, time management, and flexible thinking. Individuals with ASD often struggle with executive functioning, which can affect their ability to initiate and complete tasks, transition between activities, and adapt to changes in routines or plans.

4. **Repetitive Behaviors:** Repetitive behaviors, or restricted and repetitive interests, are commonly observed in individuals with ASD. These behaviors can manifest as repetitive body movements (e.g., hand flapping), insistence on sameness, or intense fixations on specific topics. While these behaviors can provide comfort and a sense of predictability, they may limit flexibility and social interactions.

5. **Sensory Seeking:** In contrast to sensory sensitivities, some individuals with ASD actively seek out sensory stimulation. They may engage in behaviors such as rocking, spinning, or seeking intense sensory experiences. While these behaviors can provide comfort or stimulation, they may also be disruptive in certain settings.

Support Strategies

To support individuals with ASD effectively, it is essential to consider their strengths and challenges. Here are some strategies that can foster their success and well-being:

1. **Capitalizing on Strengths:** Identifying and capitalizing on the strengths of individuals with ASD can enhance their engagement and achievements. Providing opportunities for them to explore their interests and talents can boost confidence and motivation.

2. **Visual Supports:** The visual thinking abilities of individuals with ASD can be leveraged by using visual supports, such as visual schedules, social stories, or visual aids. These tools can aid in communication, promote understanding, and reduce anxiety.

3. **Structured Environments:** Creating structured and predictable environments can help individuals with ASD navigate daily challenges. Clear routines, visual timetables, and consistent expectations can provide a sense of security and reduce stress.

4. **Social Skills Training:** Implementing social skills training programs can help individuals with ASD develop and improve their social communication skills. These programs focus on teaching specific social cues, conversation skills, and perspective-taking.

5. **Sensory Supports:** Recognizing and accommodating sensory sensitivities or sensory seeking behaviors is crucial for individuals with ASD. Providing sensory breaks, noise-canceling headphones, or sensory-friendly environments can help mitigate sensory challenges.

6. **Collaboration and Advocacy:** Collaboration among educators, families, and professionals is crucial for creating inclusive environments and supporting individuals with ASD. Engaging in advocacy efforts can promote understanding, reduce stigma, and advocate for inclusive policies and practices.

Real-World Example: Temple Grandin

Temple Grandin, a renowned professor and autism advocate, exemplifies the strengths of individuals with ASD. Despite facing challenges in social communication, she leveraged her visual thinking skills to revolutionize animal welfare practices in the livestock industry. Grandin's unique perspective allowed her to design and implement more humane livestock handling systems, benefiting both animals and the industry as a whole. Her success serves as an inspiration for individuals with ASD and highlights the importance of recognizing and embracing their strengths.

Exercises

1. Describe a real-world scenario where an individual with ASD's attention to detail could be a valuable asset.

2. Develop a visual support tool, such as a visual schedule, to assist an individual with ASD in their daily routine.

3. Imagine you are a teacher in an inclusive classroom. How would you modify your teaching strategies to accommodate the sensory sensitivities of a student with ASD?

4. Research a successful entrepreneur with ASD and identify the specific strengths that contribute to their achievements.

5. Role-play a social interaction scenario with a friend, implementing strategies from a social skills training program to enhance communication and understanding.

Further Resources

1. Book: "Neurotribes: The Legacy of Autism and the Future of Neurodiversity" by Steve Silberman.

2. TED Talk: "The World Needs All Kinds of Minds" by Temple Grandin.

3. Website: Autism Speaks (www.autismspeaks.org) - Provides comprehensive information on autism, including resources for support and advocacy.

4. Documentary: "Life, Animated" - Explores the life of Owen Suskind, an individual with autism, and highlights the power of visual storytelling and self-expression.

5. Research Paper: "Strengths of Autism: Cognitive and Behavioral Features" by Michelle Dawson and Morton Gernsbacher.

Neurodivergent Narratives in ASD

One of the most powerful aspects of the neurodiversity movement is the promotion of neurodivergent narratives. These narratives allow individuals with Autism Spectrum Disorder (ASD) to share their unique experiences, challenges, and successes, providing valuable insights into the diverse world of autism.

Neurodivergent narratives in ASD play a crucial role in challenging misconceptions and stereotypes surrounding autism. They provide a platform for individuals with ASD to reclaim their voice, showcasing their strengths, talents, and perspectives. By sharing their stories, they contribute to a more accurate and nuanced understanding of autism, breaking down the barriers of stigma and discrimination.

It is important to recognize that neurodivergent narratives are not limited to personal accounts; they also include artistic expressions such as literature, music, and visual arts. These creative outlets allow individuals with ASD to explore and communicate their experiences in unique and meaningful ways.

In addition to challenging stereotypes, neurodivergent narratives also promote acceptance and inclusion. When individuals with ASD have the opportunity to share their stories, it creates empathy and understanding among neurotypical individuals, fostering a more compassionate and inclusive society. These narratives humanize the experiences of individuals with ASD, reminding us that behind every diagnosis is a person with their own aspirations, dreams, and talents.

Neurodivergent narratives also provide a sense of community and support for individuals with ASD. When individuals share their stories, they create connections and find solace in knowing that they are not alone. These narratives can serve as a source of inspiration and empowerment, encouraging individuals with ASD to embrace their uniqueness, build self-confidence, and advocate for their needs and rights.

Furthermore, neurodivergent narratives contribute to research and education in the field of autism. By sharing their experiences, individuals with ASD provide valuable insights that can inform the development of effective interventions, supports, and policies. These narratives challenge researchers and professionals to move beyond a deficit-based approach and adopt a strengths-based perspective, focusing on the abilities and potential of individuals with ASD.

To encourage the sharing of neurodivergent narratives in ASD, it is important to create safe and inclusive spaces where individuals feel comfortable expressing themselves. Schools, workplaces, and community organizations can play a vital role in providing platforms for individuals with ASD to share their stories. These platforms can include storytelling events, art exhibitions, online blogs, and social media campaigns.

In conclusion, neurodivergent narratives in ASD are a powerful tool for promoting acceptance, understanding, and inclusion. They challenge misconceptions, provide support and empowerment, and contribute to research and education in the field of autism. By amplifying the voices of individuals with ASD, we can create a society that celebrates the diversity and richness of human experiences. Let us listen to these narratives, learn from them, and celebrate the unique strengths and talents of individuals with ASD.

Supporting Individuals with ASD in Educational and Work Settings

In order to create inclusive and supportive environments for individuals with Autism Spectrum Disorder (ASD), it is important to consider their unique strengths, challenges, and individual needs. This section aims to provide strategies and guidelines for supporting individuals with ASD in both educational and work settings. By implementing appropriate accommodations and fostering a culture of acceptance, we can empower individuals with ASD to thrive and reach their full potential.

Understanding the Needs of Individuals with ASD

Before discussing specific strategies, it is essential to have a good understanding of the needs and characteristics of individuals with ASD. Autism is a neurodevelopmental disorder characterized by challenges in social communication and interaction, as well as restricted and repetitive behaviors or interests. Individuals with ASD may also present with sensory sensitivities, difficulties with executive functioning, and specific pattern of strengths and weaknesses.

It is important to recognize that individuals with ASD have diverse strengths and talents, including attention to detail, visual thinking, and the ability to focus deeply on particular interests. However, they may also face challenges in areas such as social skills, communication, and executive functions like organization and task initiation.

Creating an Autism-Friendly Learning Environment

In an educational setting, there are several strategies that can support individuals with ASD and enhance their learning experience:

- **Individualized Education Plans (IEPs):** Collaborate with parents, teachers, and professionals to develop IEPs that address the specific needs and goals of the student with ASD. These plans should outline accommodations, modifications, and specialized instruction that will support their learning.
- **Visual Supports:** Utilize visual supports such as schedules, social stories, and visual cues to enhance understanding and promote independence. Visual supports can help individuals with ASD navigate daily routines, understand expectations, and improve their communication skills.
- **Structured Environment:** Establish a structured and predictable environment to help individuals with ASD feel secure and reduce anxiety. This can be achieved through clear routines, visual schedules, and consistent expectations.
- **Social Skills Training:** Provide explicit instruction and opportunities for individuals with ASD to develop social skills. This can be done through social skills groups, role-playing exercises, and targeted interventions focusing on effective communication, perspective-taking, and relationship-building.
- **Sensory Considerations:** Understand and address sensory sensitivities and challenges that individuals with ASD may experience. This can involve creating sensory-friendly spaces, providing noise-canceling headphones, or offering breaks to accommodate sensory needs.

By implementing these strategies, educators can create an inclusive and supportive learning environment that meets the unique needs of individuals with ASD.

Supporting Individuals with ASD in the Workplace

Transitioning from school to the workplace can be challenging for individuals with ASD. However, with the right support and accommodations, they can excel in various job roles. Here are some strategies for supporting individuals with ASD in the workplace:

- **Employer Education and Awareness:** Educate employers about the strengths and challenges of individuals with ASD, as well as the benefits of neurodiversity in the workplace. Raising awareness can help employers better understand how to support and accommodate employees with ASD.

- **Job Coaching and Mentoring:** Provide job coaches or mentors to support individuals with ASD in navigating workplace expectations, learning job tasks, and building social skills. Job coaches can also help facilitate communication between the employee and their colleagues, as well as provide ongoing support and feedback.

- **Workplace Accommodations:** Implement reasonable accommodations in the workplace to address the specific needs of individuals with ASD. Accommodations may include providing a quiet workspace, using visual supports, or adjusting work schedules to accommodate sensory sensitivities or executive functioning challenges.

- **Social Skills Training:** Offer social skills training or workshops designed to enhance interpersonal skills, communication, and problem-solving abilities in the workplace. These training programs can empower individuals with ASD to navigate social interactions and build positive relationships with their colleagues.

- **Employee Resource Groups (ERGs):** Establish ERGs or affinity groups within the organization to create a sense of community and peer support among employees with ASD. ERGs can serve as a platform for sharing experiences, providing resources, and fostering a supportive workplace culture.

By implementing these strategies, employers can create a supportive and inclusive work environment that recognizes and harnesses the unique talents of individuals with ASD.

Real-World Examples

Let's take a look at two real-world examples that highlight successful strategies for supporting individuals with ASD in educational and work settings:

- **Case Study 1: Inclusive Classroom**

  In a middle school classroom, a student with ASD was struggling with transitions and completing assignments. The teacher implemented a visual schedule, outlining the daily tasks and providing clear expectations. The student also received individualized support from a special education teacher, who provided additional structure, cues, and reminders. As a result, the student's anxiety decreased, and their academic performance improved.

- **Case Study 2: Neurodiversity Hiring Program**

  Company XYZ implemented a neurodiversity hiring program that specifically aimed to recruit individuals with ASD. They partnered with autism advocacy organizations to provide training, support, and job-coaching services for the hired employees. The program focused on leveraging the strengths of individuals with ASD, such as attention to detail, pattern recognition, and analytical thinking. As a result, the company saw increased productivity, improved problem-solving abilities, and a higher employee retention rate among individuals with ASD.

These examples illustrate the positive impact that appropriate support and accommodations can have on individuals with ASD, both in educational and work settings.

Conclusion

Supporting individuals with ASD in educational and work settings requires a proactive and inclusive approach that recognizes their unique strengths and challenges. By creating an autism-friendly learning environment, providing appropriate accommodations, and fostering acceptance and understanding, we can empower individuals with ASD to achieve success and thrive in their educational journey and careers. It is crucial to continue advocating for neurodiversity in these settings and ensure that individuals with ASD have equal opportunities to reach their full potential.

Advocacy and Activism within the Autism Community

Advocacy and activism play a crucial role in promoting acceptance, understanding, and inclusion of individuals on the autism spectrum. The autism community has a long history of grassroots movements and organized campaigns to raise awareness about autism and to fight for the rights and well-being of autistic individuals. This section will explore the different forms of advocacy and activism within the autism community and the impact they have had on shaping public perception, policies, and practices.

History of Autism Advocacy

Autism advocacy has evolved over time, reflecting changes in societal attitudes towards autism and the recognition of autistic rights. In the early days, parents and professionals were the primary advocates for autistic individuals, primarily focusing on finding a cure or treatment for autism. However, as the neurodiversity movement gained momentum in the late 1990s, the focus shifted towards celebrating autism as a natural variation of human diversity rather than a disorder to be fixed. This shift in perspective laid the foundation for self-advocacy, where autistic individuals themselves became active participants in advocating for their rights.

Self-Advocacy and Empowerment

Self-advocacy is a powerful form of activism within the autism community, as it gives voice to the experiences and perspectives of autistic individuals. By promoting self-advocacy, autistic individuals are empowered to challenge stereotypes, misinformation, and discrimination. They play a pivotal role in educating society and dispelling myths about autism, fostering acceptance, and embracing neurodiversity.

Autism self-advocates use various platforms to amplify their voices, including social media, blogs, and public speaking engagements. They share personal stories, advocacy initiatives, and educate the public about the strengths, challenges, and unique experiences of being autistic. Through self-advocacy, autistic individuals actively contribute to policy development, research, and decision-making processes that affect their lives.

Community Organizations and Support Groups

Numerous community organizations and support groups have emerged to provide a collective voice for autistic individuals and their families. These organizations aim to create a supportive network, share resources, and advocate for change at the local, national, and international levels.

Autism Speaks is one of the largest autism advocacy organizations globally. It focuses on promoting research, raising awareness, and providing resources for families. However, it has also been criticized by some autistic self-advocates for its approach, which they argue promotes fear and stigmatization of autism.

In contrast, organizations like the Autistic Self Advocacy Network (ASAN) prioritize autistic self-advocacy and work towards creating more inclusive and accessible environments for autistic individuals. ASAN actively engages in policy advocacy, community organizing, and provides resources and support for autistic individuals and their families.

Legislative Advocacy and Policy Change

Legislative advocacy involves influencing government policies, laws, and regulations to protect the rights and improve the lives of autistic individuals. Autism advocacy organizations often play a significant role in lobbying for legislative changes that promote inclusion, access to education, healthcare, employment, and other essential services.

For example, the passage of the Autism Collaboration, Accountability, Research, Education, and Support (CARES) Act in the United States in 2019 demonstrated the impact of legislative advocacy. This legislation aimed to allocate resources towards autism research, early intervention, and support services.

However, legislative advocacy also faces challenges. Autistic individuals and their allies must navigate complex political landscapes, social stigma, and limited funding to achieve meaningful policy change. It requires collaboration, persistence, and effective communication to influence decision-makers and drive systemic change.

Awareness Campaigns and Public Education

Awareness campaigns have been instrumental in increasing public understanding of autism and challenging prevailing stereotypes. These campaigns aim to promote acceptance and inclusion by educating the public on the diverse abilities, strengths, and experiences of autistic individuals.

One prominent example is "Autism Acceptance Month" celebrated every April, which seeks to reframe autism awareness as accepting and embracing neurodiversity. This month serves as an opportunity to raise public awareness through social media campaigns, community events, and educational initiatives.

Public education plays a crucial role in dismantling misconceptions and reducing stigma associated with autism. Activists and advocates organize workshops, conferences, and training sessions to educate professionals, educators, and the general public about autism. By dispelling myths and providing accurate information, these educational initiatives contribute to creating more inclusive and supportive environments for autistic individuals.

Impact and Future Directions

Advocacy and activism within the autism community have achieved significant milestones in promoting acceptance and understanding. However, challenges persist, and there is still much work to be done to ensure full inclusion for autistic individuals.

The future of autism advocacy lies in strengthening alliances with other social justice movements, such as disability rights, LGBTQ+ rights, and racial justice, to address intersecting forms of discrimination and oppression. Collaborative efforts can amplify the voices of autistic individuals and advocate for comprehensive policy changes that address the unique needs and challenges they face.

Furthermore, technology has a crucial role to play in advancing autism advocacy. Online platforms and social media provide spaces for autistic self-advocates to connect, share resources, raise awareness, and influence public opinion. Innovations in assistive technology and accessibility also contribute to creating more inclusive environments where autistic individuals can thrive.

By continuing to promote advocacy and activism, the autism community can drive meaningful change in society, fostering acceptance, inclusion, and celebrating the rich diversity of neurodivergent individuals.

Conclusion

Advocacy and activism are powerful tools in promoting acceptance, understanding, and inclusion within the autism community. From self-advocacy to community organizations, legislative advocacy to awareness campaigns, these efforts have brought about significant changes in societal attitudes, policies, and practices.

The future holds great potential for further advancements in autism advocacy. It requires a collective effort, collaboration, and ongoing education to create a world that embraces neurodiversity and supports the rights and well-being of autistic individuals. By amplifying the voices of autistic individuals and challenging systemic barriers, we can create a more inclusive and accepting society for everyone.

Attention-Deficit/Hyperactivity Disorder (ADHD)

Definition and Diagnosis of ADHD

Attention-Deficit/Hyperactivity Disorder (ADHD) is a neurodevelopmental disorder that affects both children and adults. It is characterized by persistent and impairing symptoms of inattention, hyperactivity, and impulsivity. ADHD can have a significant impact on various aspects of an individual's life, including academic performance, social relationships, and daily functioning.

Defining ADHD

ADHD is a complex condition that involves a dysregulation of brain functions related to attention, impulse control, and executive functioning. Individuals with ADHD often struggle to sustain attention, filter out distractions, and regulate their impulses. They may exhibit a range of behaviors that disrupt their ability to perform tasks, follow instructions, and interact with others in a socially appropriate manner.

The Diagnostic and Statistical Manual of Mental Disorders (DSM-5) outlines the criteria for diagnosing ADHD. According to the DSM-5, there are three subtypes of ADHD:

1. Inattentive presentation: Individuals with this subtype primarily struggle with focus, organization, and completing tasks. They may appear dreamy or forgetful and have difficulty following instructions or listening attentively.

2. Hyperactive/impulsive presentation: Individuals with this subtype exhibit high levels of activity, fidgeting, and restlessness. They may have difficulty waiting their turn, interrupting others, and engaging in impulsive behaviors without considering the consequences.

3. Combined presentation: Individuals with this subtype display symptoms of both inattention and hyperactivity/impulsivity.

Diagnosis of ADHD

Accurate diagnosis of ADHD is essential to ensure appropriate interventions and supports for individuals with the condition. The diagnostic process typically involves gathering information from multiple sources, including the individual, family members, teachers, and other professionals.

To diagnose ADHD, the following steps are typically taken:

1. Clinical interview: A healthcare professional will conduct a comprehensive interview with the individual and their caregivers to gather information about the person's symptoms, development, and functioning in different environments.

2. Rating scales and questionnaires: Standardized rating scales, such as the ADHD Rating Scale, are often used to assess the severity and frequency of ADHD symptoms. These questionnaires provide valuable information about the presence of ADHD symptoms in different settings, such as at home and school.

3. Medical evaluation: It is important to rule out any medical conditions that may mimic or contribute to ADHD symptoms. A healthcare provider may conduct a physical examination and review the individual's medical history to exclude other conditions that may be causing or exacerbating the symptoms.

4. Psychoeducational assessment: This assessment involves cognitive and academic testing to evaluate the individual's intellectual abilities, learning strengths and weaknesses, and academic performance. It helps identify any coexisting learning disorders or difficulties that may be present alongside ADHD.

5. Observation: In some cases, direct observation of the individual's behavior in different settings, such as a classroom or home, may be necessary to gain a better understanding of their symptoms and functional impairments.

It is important to note that the diagnosis of ADHD should be made by a qualified healthcare professional, such as a psychiatrist, psychologist, or pediatrician, who specializes in neurodevelopmental disorders. The professional will consider the duration, frequency, and intensity of symptoms, as well as the impact of these symptoms on the individual's daily life, before making a diagnosis.

Controversies and Challenges

Diagnosing ADHD can be challenging due to several factors. Firstly, symptoms of ADHD can vary widely among individuals, and they can also overlap with symptoms of other conditions, such as anxiety or depression. This highlights the importance of conducting a thorough assessment and considering comorbidities.

Secondly, there is ongoing debate regarding the appropriate age for diagnosing ADHD. While the DSM-5 provides criteria for diagnosing ADHD in children as young as age 4, some experts argue that it may not be developmentally appropriate to diagnose ADHD in younger children, as some of the symptoms may be part of normal childhood development.

Additionally, cultural and contextual factors can influence the interpretation of ADHD symptoms. Behaviors that are considered problematic in one culture may be more acceptable or normative in another. It is crucial to consider cultural diversity and avoid misdiagnosis or overdiagnosis of ADHD based on cultural biases or expectations.

Intervention and Treatment

Once a diagnosis of ADHD is made, appropriate interventions and treatment strategies can be implemented. The goal of treatment is to reduce the symptoms of ADHD and improve the individual's overall functioning and quality of life. Treatment approaches may include:

1. Behavioral interventions: These interventions focus on promoting positive behaviors, improving self-control, and developing organizational and

time-management skills. Behavioral strategies may include reward systems, behavior contracts, and teaching self-monitoring techniques.

2. Educational support: Individuals with ADHD often benefit from academic accommodations and support in the classroom, such as preferential seating, additional time for assignments or tests, and the use of assistive technologies.

3. Medication: In some cases, medication may be prescribed to help manage ADHD symptoms. Stimulant medications, such as methylphenidate and amphetamines, are commonly used and have been shown to be effective in reducing symptoms of ADHD. However, medication should always be used as part of a comprehensive treatment plan and under the supervision of a healthcare professional.

4. Parent and family support: Providing education and support to parents and family members can be beneficial in managing ADHD. Parent training programs that teach behavioral strategies and coping skills can help improve the parent-child relationship and reduce the impact of ADHD on family dynamics.

5. Psychotherapy: Therapy can help individuals with ADHD develop coping strategies, improve self-esteem, and address any coexisting mental health concerns. Cognitive-behavioral therapy (CBT) and social skills training are commonly used approaches for individuals with ADHD.

It is important to note that each individual with ADHD is unique, and interventions should be tailored to their specific needs and circumstances. A multidisciplinary approach involving collaboration between healthcare providers, educators, and families is crucial for successful management of ADHD.

Contemporary Issues and Research

Research on ADHD is ongoing, with new insights and advancements emerging regularly. Some of the current areas of focus in ADHD research include:

1. Genetics and neurobiology: Scientists are increasingly exploring the genetic and neurobiological underpinnings of ADHD to better understand its causes and develop targeted treatments.

2. Non-pharmacological interventions: Research is examining the effectiveness of various non-pharmacological interventions, such as mindfulness-based practices, neurofeedback, and exercise, in reducing ADHD symptoms and improving overall functioning.

3. Long-term outcomes: Longitudinal studies are being conducted to explore the impact of ADHD on individuals across the lifespan, including educational attainment, employment outcomes, and mental health.

4. Neurodiversity perspective: There is growing recognition of the neurodiversity paradigm, which emphasizes the acceptance and celebration of neurological differences. This perspective challenges the traditional deficit-based approach to ADHD and advocates for a strengths-based approach that acknowledges the unique abilities and talents of individuals with ADHD.

As research continues to advance our understanding of ADHD, it is important to incorporate these findings into clinical practice and promote awareness and support for individuals with ADHD in all aspects of life.

Summary

ADHD is a neurodevelopmental disorder characterized by symptoms of inattention, hyperactivity, and impulsivity. Diagnosis involves a comprehensive assessment that considers the individual's history, symptoms, and functional impairments. Treatment approaches may include behavioral interventions, educational support, medication, and therapy. Ongoing research is shedding light on the genetic and neurobiological factors underlying ADHD, as well as exploring new interventions and approaches. It is important to approach the diagnosis and management of ADHD with a holistic and individualized perspective, considering the unique strengths and needs of each individual.

Understanding the Heterogeneity of ADHD

Attention-Deficit/Hyperactivity Disorder (ADHD) is a neurodevelopmental condition characterized by persistent patterns of inattention, hyperactivity, and impulsivity that interfere with daily functioning. However, it is important to recognize that ADHD is a highly heterogeneous disorder, meaning that individuals with ADHD can display a range of symptoms, severity levels, and associated features. In this section, we will explore the various dimensions of heterogeneity in ADHD to gain a deeper understanding of this complex condition.

Symptom Presentation and Subtypes

ADHD symptoms can manifest in different ways, leading to the identification of distinct subtypes. The Diagnostic and Statistical Manual of Mental Disorders (DSM-5) recognizes three subtypes of ADHD:

1. Predominantly Inattentive Presentation: Individuals with this subtype struggle predominantly with inattention. They may have difficulty sustaining attention, organizing tasks, following instructions, and completing assignments.

2. Predominantly Hyperactive-Impulsive Presentation: Individuals with this subtype exhibit hyperactivity and impulsivity as the primary symptoms. They may be fidgety, talk excessively, have difficulty waiting their turn, and frequently interrupt others.

3. Combined Presentation: Individuals with the combined presentation experience significant symptoms of both inattention and hyperactivity-impulsivity.

It is important to note that while these subtypes provide a framework for classification, many individuals with ADHD exhibit symptoms that do not neatly fit into these categories. This highlights the diversity of symptom presentations within the ADHD population.

Impairment and Functional Outcomes

The heterogeneity of ADHD extends beyond symptom presentation to the level of impairment and functional outcomes. While some individuals with ADHD may experience significant impairment in multiple domains of life, including academic performance, occupational functioning, and interpersonal relationships, others may exhibit milder impairment and demonstrate resilience in certain areas.

Factors such as intelligence, co-occurring conditions (e.g., anxiety, depression), and the presence of supportive environments can influence the impact of ADHD on individuals' lives. For example, a highly intelligent individual with ADHD may succeed academically despite their symptoms, while another individual with average intelligence may struggle more due to additional challenges. It is crucial to consider these individual differences when assessing and providing support for individuals with ADHD.

Neurobiological Variability

The heterogeneity of ADHD also extends to its underlying neurobiology. Research suggests that ADHD is associated with alterations in the structure and functioning of various brain regions involved in attention, executive functions, and reward processing.

However, studies have demonstrated that there is significant variability in these neurobiological characteristics among individuals with ADHD. For example, some individuals may show reduced cortical thickness in certain brain regions, while others may not display such differences. Additionally, neuroimaging studies have revealed that ADHD-related brain differences can vary depending on age, sex, and the presence of comorbid conditions.

This neurobiological heterogeneity in ADHD not only contributes to the diversity in symptom presentation but also has implications for treatment approaches. Recognizing these individual differences in brain functioning can help tailor interventions to address specific needs and capitalize on strengths.

Environmental and Psychosocial Factors

In addition to neurobiological variability, environmental and psychosocial factors further contribute to the heterogeneity of ADHD. These factors can shape the expression and impact of ADHD symptoms.

For example, certain environmental risk factors, such as prenatal exposure to tobacco smoke, low birth weight, and early life adversity, have been associated with an increased likelihood of developing ADHD. Furthermore, the quality of parenting, educational settings, and the presence of supportive relationships can significantly influence how ADHD symptoms manifest and affect individuals' lives.

Understanding the interplay between individual characteristics, neurobiological factors, and environmental influences is crucial for a comprehensive understanding of ADHD's heterogeneity. It emphasizes the importance of taking a holistic and individualized approach in treatment and support interventions.

Implications for Assessment and Intervention

The heterogeneity of ADHD poses challenges for accurate assessment and effective intervention. It highlights the importance of comprehensive evaluations that consider multiple dimensions of the disorder and individual differences.

Assessment tools should capture the diverse ways in which ADHD symptoms can manifest and the impact they have on various domains of functioning. It is essential to involve multiple informants, such as parents, teachers, and the individual themselves, to gather a comprehensive understanding of symptom severity and functional impairments.

Similarly, interventions need to be tailored to address the specific needs of each individual. This may involve a combination of medication management, behavioral interventions, educational accommodations, and psychosocial support. In recognizing the heterogeneity of ADHD, it is crucial to adopt a personalized and flexible approach to treatment that considers the unique strengths and challenges of each individual.

Real-world Example: Workplace Accommodations

To illustrate the impact of heterogeneity in ADHD, let's consider a real-world example related to workplace accommodations. Two individuals with ADHD may both experience difficulties with maintaining attention, organization, and time management. However, their specific challenges may vary based on their symptoms and the nature of their work.

For instance, one individual may struggle with concentration during long meetings and require extra assistance in organizing their workload. On the other hand, another individual may experience difficulties with impulse control, leading to impulsive decision-making in high-pressure situations. In this case, they may benefit from strategies that promote self-awareness and self-regulation.

These examples highlight the need for personalized workplace accommodations that address the specific challenges individuals with ADHD may face. Tailoring support measures to the unique needs of each individual can enhance their productivity, job satisfaction, and overall well-being.

Summary

In this section, we explored the heterogeneity of ADHD from various perspectives. We discussed how symptom presentation can differ among individuals, leading to the identification of subtypes. We also recognized how ADHD can vary in terms of impairment and functional outcomes, neurobiological characteristics, and the influence of environmental and psychosocial factors.

Understanding the heterogeneity of ADHD is crucial for accurate assessment, effective intervention, and the provision of individualized support. By recognizing the diverse nature of ADHD, we can promote a more comprehensive and personalized approach to understanding and addressing this complex neurodevelopmental condition.

Strengths and Challenges of Individuals with ADHD

Individuals with Attention-Deficit/Hyperactivity Disorder (ADHD) possess a unique set of strengths and face specific challenges that can have a significant impact on various aspects of their lives. Understanding these strengths and challenges is essential for providing appropriate support and promoting the well-being of individuals with ADHD.

Strengths of Individuals with ADHD

Despite the commonly known challenges associated with ADHD, individuals with this condition possess several strengths that should not be overlooked. These strengths include:

Creativity and Innovation: Individuals with ADHD often exhibit exceptional creativity and innovative thinking. Their minds are wired to make unique connections between seemingly unrelated ideas, leading to out-of-the-box thinking and problem-solving abilities. This creative thinking can bring fresh perspectives to various domains, such as art, technology, and entrepreneurial ventures.

For example, famous filmmaker Steven Spielberg has openly discussed his ADHD diagnosis and attributes his ability to think outside the box and create imaginative films to his ADHD-related strengths.

Hyperfocus: While attention difficulties are common in individuals with ADHD, they also have the potential for hyperfocus. Hyperfocus refers to a state of intense concentration and absorption in a task that interests or engages them. During periods of hyperfocus, individuals with ADHD demonstrate exceptional productivity and a remarkable ability to tune out distractions.

For instance, a student with ADHD might struggle to stay focused during regular classroom activities but can fully engage in a project they are passionate about, demonstrating heightened focus and attention to detail.

High Energy Levels: Many individuals with ADHD possess high levels of energy. While this can sometimes be challenging to manage, it also translates into enthusiasm, vitality, and a willingness to take on new challenges. Their energy can be contagious and can inspire others around them.

Consider an individual with ADHD who brings infectious enthusiasm and motivates their team members during a group project, fueling their passion and sparking their creativity.

Challenges Faced by Individuals with ADHD

ADHD also presents individuals with specific challenges that are important to understand and address. Some of these challenges are:

Inattention and Distractibility: The hallmark symptoms of ADHD include inattention and distractibility. Individuals with ADHD often struggle to maintain focus and attention on tasks that are not stimulating or engaging to them. This can lead to difficulties in academic, occupational, and day-to-day activities that require sustained attention.

For instance, a student with ADHD may find it challenging to concentrate during lectures or while studying, resulting in decreased academic performance.

Impulsivity: Impulsivity is another core symptom of ADHD. Individuals with ADHD often act without thinking, making quick decisions without considering the consequences. This impulsivity can manifest in various ways, such as interrupting others during conversations or engaging in risky behaviors.

For example, an individual with ADHD may impulsively make impulsive purchases without considering their financial implications, leading to financial difficulties.

Time Management and Organization: Many individuals with ADHD struggle with time management and organization. They may have difficulty estimating the time required for tasks, prioritizing activities, and keeping track of deadlines. This can lead to increased stress, missed deadlines, and a sense of being overwhelmed.

Consider a professional with ADHD who frequently misses important meetings or deadlines due to difficulties with time management and organization.

Social and Interpersonal Difficulties: Individuals with ADHD commonly experience challenges in social interactions and interpersonal relationships. They may struggle with reading social cues, maintaining appropriate conversations, and controlling impulsive behaviors. These difficulties can impact their ability to form and maintain friendships and may lead to feelings of social isolation.

For instance, a child with ADHD may have difficulty waiting their turn in a conversation, often interrupting others and causing strain in their relationships.

Academic and Occupational Challenges: ADHD can significantly impact academic and occupational performance. Difficulties with attention, organization, and impulsivity can make it challenging for individuals with ADHD to meet expectations and perform at their full potential. They may struggle with completing tasks, following instructions, and staying on track with their responsibilities.

For example, an adult with ADHD may find it challenging to meet deadlines and stay organized at work, leading to decreased productivity and potential performance issues.

Emotional Regulation: Emotional regulation is often compromised in individuals with ADHD. They may have difficulty managing their emotions, resulting in heightened emotional responses, impulsivity, and mood swings. This can impact their overall well-being and relationships with others.

Consider a teenager with ADHD who struggles to control their anger outbursts, often leading to strained relationships with family members and peers.

Addressing the Challenges and Leveraging the Strengths

To support individuals with ADHD effectively, it is crucial to address the challenges they face while leveraging their strengths. Here are some strategies and considerations:

Individualized Accommodations: Providing individualized accommodations is essential to help individuals with ADHD overcome their challenges. This may include environmental modifications, such as providing a quiet workspace or minimizing distractions, as well as implementing organizational tools and time management strategies.

For instance, a student with ADHD may benefit from using a planner or digital reminders to track assignments and deadlines.

Promoting Self-Advocacy: Encouraging individuals with ADHD to advocate for themselves can empower them to communicate their needs and seek appropriate support. Teaching self-advocacy skills, such as requesting accommodations or expressing personal challenges, can promote their independence and resiliency.

Consider providing resources and guidance to individuals with ADHD to initiate conversations with teachers or supervisors about their specific needs.

Skill Building: Supporting individuals with ADHD in developing strategies to manage their challenges can significantly improve their well-being and functioning. Skill-building interventions may focus on areas such as time management, organization, emotional regulation, and social skills.

For example, teaching stress-management techniques or providing social skills training can help individuals with ADHD navigate challenges more effectively.

Strength-Based Approaches: Recognizing and nurturing the strengths of individuals with ADHD can boost their self-confidence, motivation, and overall success. Creating opportunities that allow them to showcase their creativity, engage in hyperfocus, and channel their energy positively can lead to remarkable accomplishments.

Consider involving individuals with ADHD in projects or activities that harness their creative thinking or provide the opportunity for hyperfocus.

In conclusion, individuals with ADHD possess unique strengths, such as creativity, hyperfocus, and high energy levels, along with specific challenges, including inattention, impulsivity, and difficulties with time management. By understanding these strengths and challenges, and implementing appropriate strategies to address them, we can create a supportive environment that celebrates the neurodiversity of individuals with ADHD.

Strategies for Supporting Individuals with ADHD

Supporting individuals with Attention-Deficit/Hyperactivity Disorder (ADHD) is crucial for their success and well-being. It requires a comprehensive approach that addresses both the academic and social-emotional aspects of their lives. This section will discuss strategies that can be implemented to support individuals with ADHD in various settings, including educational institutions and workplaces.

Understanding the Needs of Individuals with ADHD

Before discussing specific strategies, it is essential to have a solid understanding of the needs and challenges faced by individuals with ADHD. ADHD is a neurodevelopmental disorder characterized by persistent patterns of inattention, impulsivity, and hyperactivity that significantly impact daily functioning. It affects people of all ages and can have a profound impact on their academic performance, social interactions, and overall well-being.

Individuals with ADHD may struggle with executive functioning skills, such as organization, time management, and planning. They may have difficulty focusing and sustaining attention, leading to poor academic performance and incomplete assignments. Additionally, impulsivity and hyperactivity can make it challenging for individuals with ADHD to regulate their behavior and emotions, leading to social difficulties and disrupted relationships.

Understanding the specific needs of individuals with ADHD allows educators, employers, and support networks to tailor strategies that address their unique challenges effectively.

Multimodal Treatment Approach

The treatment for ADHD typically involves a multimodal approach that combines different strategies to provide comprehensive support. This approach may include a combination of medication, behavioral interventions, environmental modifications, and educational accommodations.

1. Medication: The use of stimulant or non-stimulant medications, prescribed by healthcare professionals, can significantly reduce the symptoms of ADHD and improve attention, focus, and impulse control. It is essential for individuals with ADHD to work closely with their healthcare providers to find the right medication and dosage that suits their needs.

2. Behavioral Interventions: Behavioral interventions aim to teach individuals with ADHD specific skills and strategies to manage their symptoms effectively. These interventions may include:

- Behavior therapy: It focuses on teaching individuals skills such as self-monitoring, problem-solving, and self-reinforcement. Through positive reinforcement, individuals learn to replace impulsive and disruptive behaviors with more appropriate alternatives.

- Cognitive-behavioral therapy (CBT): CBT helps individuals with ADHD recognize and challenge negative thinking patterns that contribute to their difficulties. It also teaches them coping strategies to handle stress, improve time management, and strengthen organizational skills.

- Social skills training: This intervention focuses on teaching individuals with ADHD appropriate social behaviors, such as active listening, empathy, and conflict resolution. It helps improve their social interactions and build positive relationships.

3. Environmental Modifications: Modifying the environment of individuals with ADHD can create a supportive and conducive setting. Some strategies include:

- Providing a structured and predictable routine: Establishing clear routines and schedules can help individuals with ADHD manage their time effectively and reduce anxiety.

- Using visual cues and reminders: Visual supports, such as calendars, to-do lists, and visual schedules, can assist individuals with ADHD in staying organized and on track.

- Creating an organized workspace: Removing distractions and clutter from the workspace can help individuals with ADHD maintain focus and minimize sensory overload.

4. Educational Accommodations: In educational settings, accommodations are essential to support individuals with ADHD and help them succeed academically.

Some possible accommodations include:

- Extended time on exams and assignments: Allowing individuals with ADHD additional time can help mitigate their difficulties with time management and processing speed.
- Preferential seating: Placing individuals with ADHD in proximity to the teacher or in areas with fewer distractions can help them maintain attention during instruction.
- Breaks and movement opportunities: Allowing individuals with ADHD to take short breaks or engage in movement activities can help them release excess energy and improve focus.
- Use of assistive technology: Technology tools such as text-to-speech software, organizational apps, and digital reminders can assist individuals with ADHD in managing their assignments and staying organized.

Promoting Self-Advocacy and Self-Regulation

Empowering individuals with ADHD to become self-advocates and develop self-regulation skills is crucial for their long-term success. Here are strategies to promote self-advocacy and self-regulation:

1. Educate individuals about ADHD: Providing accurate and age-appropriate information about ADHD helps individuals understand their strengths and challenges. It promotes self-acceptance and encourages them to seek appropriate support when needed.
2. Teach self-awareness: Help individuals with ADHD recognize their symptoms, triggers, and personal coping strategies. Encourage self-reflection and the development of strategies that work best for them individually.
3. Build organizational and time management skills: Teach individuals strategies such as using planners, breaking tasks into smaller steps, and setting achievable goals. These skills help improve their productivity and reduce feelings of overwhelm.
4. Foster emotional regulation: Work with individuals to identify and express their emotions in healthy ways. Teach techniques such as deep breathing exercises, mindfulness, and problem-solving to manage stress and regulate emotional responses.
5. Encourage self-advocacy: Teach individuals how to communicate their needs effectively with teachers, employers, and peers. Encourage them to seek clarification, request accommodations, and express their strengths and challenges without fear or shame.

Collaborative Approach

Supporting individuals with ADHD requires a collaborative approach involving educators, parents, healthcare professionals, and other support networks. Collaboration enhances the consistency and effectiveness of interventions across various settings. Here are some strategies for fostering collaboration:

1. Open communication: Establish open lines of communication between educators, parents, and healthcare professionals. Regularly share updates and progress reports to ensure that everyone is informed and involved in supporting the individual with ADHD.

2. Individualized support plans: Develop and implement individualized support plans that address the unique needs and goals of each individual with ADHD. Ensure that all stakeholders are actively involved in the planning process.

3. Regular team meetings: Schedule regular team meetings to discuss progress, adjustments, and any concerns. These meetings provide an opportunity for collaboration and problem-solving.

4. Professional development: Provide ongoing professional development for educators and employers to enhance their understanding of ADHD and effective strategies for support. This training should address evidence-based practices, accommodations, and the promotion of neurodiversity.

Real-World Examples and Resources

To further illustrate the strategies discussed, here are a few real-world examples:

1. Using a digital planner and reminder app: An individual with ADHD can utilize a smartphone app that sends reminders for tasks, appointments, and deadlines. This helps them stay organized and manage their time effectively.

2. Implementing a fidget tool: In an educational setting, providing a stress ball or other discreet fidget tool can allow students with ADHD to channel their excess energy and improve focus during lessons.

3. Collaborative problem-solving: During a team meeting, educators, parents, and the individual with ADHD can collectively brainstorm strategies to address difficulties with completing assignments. This collaboration ensures that everyone's perspectives and expertise are considered.

Resources for further support and information:

- CHADD (Children and Adults with Attention-Deficit/Hyperactivity Disorder): A national nonprofit organization that provides education, advocacy, and support for individuals with ADHD. (Website: chadd.org)

- Understood: An online resource that offers practical strategies, support, and personalized resources for individuals with learning and attention issues. (Website: understood.org)

- National Resource Center on ADHD: A program of CHADD that provides comprehensive information, resources, and access to local support groups. (Website: help4adhd.org)

- The ADHD Dude: A YouTube channel and resource website created by a licensed clinical social worker that offers practical advice and strategies for individuals with ADHD. (Website: adhddude.com)

In conclusion, supporting individuals with ADHD requires a multimodal approach that addresses their unique needs and challenges. By implementing strategies such as medication, behavioral interventions, environmental modifications, and educational accommodations, individuals with ADHD can thrive academically and socially. Promoting self-advocacy, self-regulation, and collaboration across various settings further enhances their success. With the right support, individuals with ADHD can develop their strengths, overcome challenges, and lead fulfilling lives.

Emotional Regulation and ADHD

Emotional regulation refers to the ability to manage and control one's emotions effectively. It involves recognizing and understanding emotions, expressing them in a socially acceptable manner, and adapting emotionally to different situations. In the context of ADHD, emotional regulation can be challenging for individuals due to the neurobiological differences associated with the condition.

1. Understanding Emotional Regulation and ADHD:

Emotional dysregulation is commonly observed in individuals with ADHD. They may experience difficulties in managing their emotions and may exhibit intense emotional reactions, impulsivity, and difficulty in self-soothing. These challenges can manifest in various ways, such as frequent mood swings, anger outbursts, and difficulty in handling frustration or disappointment.

2. Factors contributing to Emotional Regulation Difficulties:

Several factors contribute to emotional regulation difficulties in individuals with ADHD. These factors include:

2.1. Executive function deficits: ADHD is associated with executive function deficits, which involve higher-order cognitive processes responsible for self-regulation and emotional control. These deficits can make it harder for individuals with ADHD to modulate their emotions effectively.

2.2. Emotional impulsivity: Impulsivity is a hallmark symptom of ADHD. It can result in impulsive emotional reactions, causing individuals with ADHD to act impulsively based on their emotions without considering the consequences.

2.3. Sensory sensitivity: Many individuals with ADHD have heightened sensory sensitivity, which can trigger emotional reactions. They may become overwhelmed by sensory stimuli, leading to emotional dysregulation.

2.4. Co-occurring conditions: ADHD often co-occurs with other mental health conditions, such as anxiety or depression, which can exacerbate emotional regulation difficulties.

3. Strategies for Improving Emotional Regulation:

Effective emotional regulation strategies can help individuals with ADHD manage their emotions more effectively. These strategies include:

3.1. Self-awareness: Encouraging individuals with ADHD to develop self-awareness about their emotions is an essential first step. They can learn to recognize their emotional states, identify triggers, and understand how their emotions influence their behavior.

3.2. Cognitive-behavioral techniques: Cognitive-behavioral techniques, such as cognitive restructuring and problem-solving, can help individuals with ADHD challenge negative thinking patterns and develop adaptive coping strategies.

3.3. Mindfulness and relaxation techniques: Mindfulness practices, such as deep breathing exercises and meditation, can enhance emotional self-regulation. These techniques promote self-awareness, reduce impulsivity, and improve emotional well-being.

3.4. Social support: Building a strong support network is crucial for individuals with ADHD. Having understanding and supportive people in their lives can provide emotional validation, guidance, and assistance in managing difficult emotions.

3.5. Structured routines and environments: Establishing structured routines and creating an organized environment can help individuals with ADHD regulate their emotions more effectively. Predictability and clear expectations can reduce anxiety and promote emotional stability.

4. The Role of Therapy and Medication:

In some cases, individuals with ADHD may benefit from therapy and medication to support their emotional regulation. Cognitive-behavioral therapy (CBT) can provide coping strategies and tools to manage emotions, while medication can help regulate brain chemicals associated with emotional dysregulation.

5. Real-World Example: Emotion-Focused Therapy

Emotion-Focused Therapy (EFT) is an evidence-based approach that can be beneficial for individuals with ADHD and emotional dysregulation. EFT helps individuals explore and process their emotions in therapy sessions, teaching them to regulate their emotions effectively. Through techniques such as emotion awareness and expression, individuals develop a greater understanding of their emotions and learn healthy ways to manage them.

In conclusion, emotional regulation can be challenging for individuals with ADHD due to several factors, including executive function deficits and emotional impulsivity. However, with awareness, strategies, and support, individuals with ADHD can develop effective emotional regulation skills. Therapy and medication can also play a role in supporting emotional regulation in individuals with ADHD. Embracing and understanding the emotional experiences of individuals with ADHD can promote acceptance and foster a neurodiverse society.

Dyslexia

Definition and Diagnosis of Dyslexia

Dyslexia is a specific learning disorder that affects reading, writing, and spelling abilities. It is characterized by difficulties in recognizing and decoding words, as well as problems with fluency and comprehension. Dyslexia is estimated to affect approximately 5-10

Understanding Dyslexia

Dyslexia is a neurodevelopmental condition that is believed to have a genetic basis. It is not related to intelligence or vision problems, but rather to differences in the way the brain processes language. Individuals with dyslexia often have difficulty mapping letters and sounds efficiently, which can impede their ability to decode words and read fluently.

Dyslexia can manifest in various ways and can vary in severity. Some individuals may have mild dyslexia, which only becomes noticeable when they encounter complex texts or unfamiliar words. Others may have more severe dyslexia, which can significantly impact their academic performance and daily life.

Diagnosis of Dyslexia

Diagnosing dyslexia involves a comprehensive assessment conducted by a team of professionals, including educators, psychologists, and speech-language pathologists. The evaluation typically consists of several components, which may include:

1. **Interviews and questionnaires:** The individual and their parents or caregivers will be asked about their developmental history, learning difficulties, and family history of dyslexia.

2. **Cognitive assessment:** This involves evaluating the individual's intellectual abilities, such as verbal and non-verbal reasoning, working memory, and processing speed. It helps identify strengths and weaknesses in cognitive functioning.

3. **Reading and writing assessments:** Various tests and tasks are administered to assess the individual's reading and writing skills. These may include measures of word recognition, reading comprehension, spelling, and writing fluency.

4. **Phonological awareness assessment:** Phonological awareness refers to the ability to recognize and manipulate the sounds of language. This assessment helps identify difficulties in phonological processing, a core deficit in dyslexia.

5. **Observations and classroom performance:** Observations in different settings, such as the classroom, can provide valuable insights into the individual's learning difficulties and how they cope with academic demands.

It is important to note that diagnosing dyslexia requires ruling out other possible causes of reading difficulties, such as visual or hearing impairments, intellectual disabilities, or inadequate educational opportunities. The assessment process should be thorough and consider the individual's unique strengths and challenges.

Interventions and Support

Once a diagnosis of dyslexia is made, appropriate interventions and support can be implemented to help individuals with dyslexia succeed academically and develop effective reading strategies. Some intervention approaches include:

1. **Structured literacy instruction:** This approach focuses on teaching phonological awareness, phonics, vocabulary, fluency, and comprehension skills systematically and explicitly.

2. **Multisensory techniques:** Using multisensory instructional materials and strategies can help individuals with dyslexia engage multiple senses (visual, auditory, kinesthetic) to reinforce learning.

3. **Assistive technology:** Assistive technology tools, such as text-to-speech software, word prediction, and audiobooks, can support reading and writing tasks and promote independence.

4. **Individualized education plans (IEPs):** Collaborating with educators, parents, and specialists, an IEP can be developed to outline specific goals, accommodations, and modifications to support the individual's unique needs.

5. **Self-advocacy skills:** Teaching self-advocacy skills can empower individuals with dyslexia to understand their strengths and challenges, communicate their needs, and seek appropriate accommodations or modifications.

The support and interventions provided to individuals with dyslexia should be evidence-based, individualized, and implemented by trained professionals. Ongoing monitoring and assessment are essential to track progress and make necessary adjustments to intervention strategies.

Resources and Organizations

Numerous resources and organizations provide valuable information and support for individuals with dyslexia and their families. Here are some examples:

1. International Dyslexia Association (IDA): A leading organization that offers resources, research, and advocacy for individuals with dyslexia. Their website provides information on assessment, intervention, and support systems.

2. Learning Disabilities Association of America (LDA): LDA offers support, resources, and advocacy for individuals with learning disabilities, including dyslexia. Their website provides information on educational rights, research, and support networks.

3. National Center for Learning Disabilities (NCLD): NCLD provides information, resources, and support for individuals with learning disabilities, including dyslexia. Their website offers articles, webinars, and tools for parents, educators, and individuals with dyslexia.

4. Understood.org: Understood.org is a comprehensive website that provides information and resources for individuals with learning and attention issues, including dyslexia. Their website offers articles, videos, and tools for parents, educators, and individuals seeking support.

It is important for individuals with dyslexia and their families to access reputable resources, connect with support networks, and advocate for appropriate interventions and accommodations to ensure their success and well-being.

Case Study: Emma's Dyslexia Journey

Emma is a 10-year-old girl who has been struggling with reading and writing since she started school. Her parents noticed that she had difficulty recognizing letters and frequently reversed or substituted them when reading or writing. Emma's teacher also observed that she struggled with basic phonics skills and understanding the connection between letters and sounds.

Concerned about Emma's difficulties, her parents sought an assessment to better understand her challenges. The assessment revealed that Emma had dyslexia, which explained her difficulties in reading and writing.

With the support of her school's special education team, Emma's parents and teachers developed an individualized education plan (IEP) to address her needs. Emma began receiving structured literacy instruction that focused on phonemic awareness, phonics, and decoding skills. She also received support from a speech-language pathologist who targeted her language and communication skills.

Over time, Emma's reading and writing skills improved with the consistent and targeted interventions. She also learned strategies to assist her in daily academic tasks, such as using assistive technology tools to access audiobooks and speech-to-text software for writing assignments.

Emma's self-esteem and confidence grew as she discovered her strengths and learned to advocate for herself. With ongoing support and accommodations, Emma successfully navigated her dyslexia journey and developed a love for reading and writing.

This case study illustrates the importance of timely and appropriate intervention in helping individuals with dyslexia overcome challenges and reach their full potential.

Exercises

1. Reflect on your own understanding of dyslexia before reading this section. How has your perspective changed? Discuss any misconceptions you may have had.

2. Imagine you are an educator. How would you create an inclusive classroom environment for students with dyslexia? Provide specific strategies and accommodations.

3. Research and identify a dyslexia-friendly font that can improve readability for individuals with dyslexia. Explain the design features that make it more accessible.

4. Interview a person with dyslexia or a parent of a child with dyslexia. Ask them about their experiences, challenges, and strategies they have found helpful. Prepare a summary of their insights.

5. Design a brochure or poster to raise awareness about dyslexia. Include information about its definition, common signs, and available resources for support.

Remember to approach these exercises with empathy and respect for the experiences of individuals with dyslexia.

Understanding the Learning Profile of Individuals with Dyslexia

Dyslexia is a neurodevelopmental condition that affects an individual's ability to read, spell, and process language. It is characterized by difficulties with phonological awareness, decoding, fluency, and comprehension. In order to support individuals with dyslexia effectively, it is crucial to have a thorough understanding of their unique learning profile.

1. Phonological Processing: Individuals with dyslexia often struggle with phonological processing, which is the ability to perceive and manipulate sounds in spoken language. This difficulty can result in challenges with phonemic awareness, the ability to recognize and manipulate individual sounds in words. For example, individuals with dyslexia may have difficulty segmenting words into syllables or identifying rhyming words. This can impact their ability to decode words accurately and fluently.

2. Reading Fluency: Dyslexia is also characterized by deficits in reading fluency, which refers to the ability to read text accurately, quickly, and with expression. Individuals with dyslexia may read slowly and inaccurately, often pausing or hesitating while reading. This can impact their comprehension of the text, as they may struggle to process information at a sufficient pace. It is important to provide explicit instruction and support to improve reading fluency in individuals with dyslexia.

3. Spelling and Writing: Individuals with dyslexia commonly exhibit difficulties with spelling and writing. They may have trouble with letter-sound correspondence, making it challenging to spell words phonetically. These difficulties can impact their overall written expression and may lead to frustration and reduced self-esteem. Instructional strategies that focus on explicit teaching of spelling rules, sight words, and word patterns can be beneficial for individuals with dyslexia.

4. Working Memory: Working memory refers to the ability to hold and manipulate information in one's mind over short periods of time. Individuals with dyslexia may have weaknesses in working memory, which can make it challenging for them to remember and process information while reading or writing. This can affect their ability to comprehend text, follow instructions, and organize their thoughts. Providing accommodations, such as breaking down tasks into smaller steps or using visual aids, can help individuals with dyslexia overcome working memory difficulties.

5. Visual Processing: Although dyslexia is primarily a language-based disorder, some individuals with dyslexia may also experience difficulties with visual processing. They may struggle with letter reversals, such as mistaking "b" for "d," or

have difficulty tracking and accurately perceiving printed text. It is important to assess and address any visual processing difficulties in order to support individuals with dyslexia effectively.

6. Strengths and Resilience: While dyslexia presents unique challenges, individuals with dyslexia often possess strengths such as creativity, problem-solving skills, and strong spatial reasoning abilities. These strengths should be recognized and celebrated, as they can contribute to their overall learning and development. It is crucial to foster a positive learning environment that promotes the strengths and resilience of individuals with dyslexia.

In understanding the learning profile of individuals with dyslexia, it is important to recognize that dyslexia is a continuum, and each individual's profile may vary. It is essential to adopt a personalized approach that takes into account the specific needs, strengths, and preferences of each individual. By doing so, we can create an inclusive and supportive educational environment that empowers individuals with dyslexia to thrive academically and personally.

Strengths and Challenges of Individuals with Dyslexia

Dyslexia is a neurodevelopmental condition that affects an individual's ability to read, write, and spell. While dyslexia can present challenges in academic and daily life, it is important to recognize that individuals with dyslexia also possess unique strengths and abilities. In this section, we will explore the strengths and challenges that individuals with dyslexia may face, and how these can be addressed and supported.

Strengths of Individuals with Dyslexia

Despite the difficulties in reading and writing, individuals with dyslexia often display remarkable strengths in other areas. These strengths can vary from person to person, but some common strengths observed in individuals with dyslexia include:

- **Strong problem-solving skills:** People with dyslexia often think in a visual and holistic manner, allowing them to approach problems from different angles and find creative solutions. They are often able to see patterns and connections that others might miss.

- **Good spatial reasoning:** Many individuals with dyslexia excel in spatial tasks, such as visualizing and manipulating objects in their minds. This ability can be an advantage in fields like architecture, engineering, and design, where visual thinking is crucial.

- **Highly developed intuition and empathy:** Dyslexic individuals often possess a heightened sense of intuition and empathy. They are acutely attuned to the feelings and needs of others, which can make them empathetic and compassionate individuals.

- **Creativity and outside-the-box thinking:** Dyslexia is often associated with a strong imagination and the ability to think outside the box. Many individuals with dyslexia excel in creative pursuits such as art, music, and writing (once they overcome the challenges associated with reading and spelling).

- **Exceptional verbal skills:** Although reading and writing may be challenging, individuals with dyslexia often have strong verbal skills. They may excel in oral communication, storytelling, and public speaking, utilizing their strong vocabulary and expressive abilities.

These strengths demonstrate that dyslexia is not a weakness or a deficit, but rather a different way of processing information.

Challenges Faced by Individuals with Dyslexia

While dyslexic individuals possess unique strengths, they also face various challenges that can impact their academic progress, career choices, and self-esteem. Some common challenges associated with dyslexia include:

- **Reading difficulties:** Individuals with dyslexia often struggle with reading accuracy, fluency, and comprehension. They may have difficulty recognizing words, decoding unfamiliar words, or understanding written text. This can impact their performance in school and their ability to access information.

- **Spelling challenges:** Dyslexia is frequently accompanied by spelling difficulties. Dyslexic individuals may have trouble organizing letters and sounds correctly and following spelling rules consistently. This can make written communication a time-consuming and frustrating process.

- **Working memory limitations:** Dyslexia can affect working memory, which is the ability to hold and manipulate information in one's mind temporarily. This can make it challenging for individuals with dyslexia to follow multi-step instructions, keep track of details, and remember information.

- **Slow processing speed:** Dyslexic individuals often process information at a slower pace than their peers. This can affect their ability to complete tasks efficiently, keep up with classroom instruction, and take tests within allotted time.

- **Impacts on self-esteem and self-confidence:** Struggling with reading, writing, and spelling can take a toll on the self-esteem and self-confidence of individuals with dyslexia. They may feel frustrated, anxious, or inadequate, especially when comparing themselves to their peers. It is essential to provide them with the support and encouragement they need to build resilience and maintain a positive self-image.

These challenges highlight the importance of early identification and appropriate interventions to help individuals with dyslexia overcome difficulties and unlock their full potential.

Supporting Individuals with Dyslexia

To support individuals with dyslexia, it is crucial to provide targeted interventions and accommodations that address their specific needs. Some strategies and resources that can help include:

- **Multisensory instruction:** Using multisensory techniques that engage multiple senses (such as seeing, hearing, and touching) can aid in the learning process for individuals with dyslexia. This approach helps reinforce phonetic awareness, letter-sound associations, and reading comprehension.

- **Structured literacy programs:** Structured literacy programs, such as the Orton-Gillingham approach, provide explicit and systematic instruction in decoding, spelling, and reading comprehension. These programs focus on teaching the underlying structure of the English language in a structured and cumulative manner.

- **Assistive technology:** The use of assistive technology can support individuals with dyslexia in various ways. Text-to-speech software, speech recognition software, electronic spell-checkers, and audiobooks can help with reading, writing, and spelling tasks. Electronic organizers and voice-to-text applications can assist with organization and note-taking.

- **Extended time for assignments and tests:** Providing additional time for assignments and tests can help dyslexic individuals overcome processing speed challenges and demonstrate their knowledge and abilities effectively.

- **Supportive and inclusive learning environments:** Creating a supportive and inclusive environment is essential for individuals with dyslexia. Teachers and peers should provide understanding, patience, and encouragement. Promoting self-advocacy skills and raising awareness about dyslexia can also contribute to a positive learning experience.

It is important to recognize that dyslexia does not define an individual's intelligence or potential. By providing appropriate support, accommodations, and understanding, we can help individuals with dyslexia thrive and contribute their unique strengths to society.

Conclusion

In conclusion, individuals with dyslexia face challenges in reading, writing, and spelling, but they also possess unique strengths and abilities. By understanding and addressing these challenges, and by providing appropriate support and accommodations, we can empower individuals with dyslexia to unlock their full potential. It is crucial to shift the focus from deficits to strengths, celebrating the diversity and contributions of dyslexic individuals in our neurotypical world.

Accommodations and Interventions for Individuals with Dyslexia

Dyslexia is a neurodevelopmental condition characterized by difficulties with reading, spelling, and decoding words. It is important to provide appropriate accommodations and interventions to support individuals with dyslexia in their educational and work settings. In this section, we will explore various strategies and approaches that can help individuals with dyslexia overcome challenges and reach their full potential.

Understanding Dyslexia Accommodations

Accommodations are adjustments made in the learning environment that enable individuals with dyslexia to access and participate in educational activities effectively. These accommodations are designed to minimize the impact of dyslexia-related difficulties on learning and allow individuals with dyslexia to demonstrate their knowledge and skills.

Some common dyslexia accommodations include:

1. Extended time on exams: Extra time allows individuals with dyslexia to process information at their own pace and reduces the pressure of time constraints.

2. Use of assistive technology: Assistive technology tools, such as text-to-speech software and speech recognition programs, can help individuals with dyslexia read and write more efficiently.

3. Multi-sensory learning materials: Utilizing materials and instructional techniques that engage different senses, such as visual, auditory, and kinesthetic, can enhance learning for individuals with dyslexia.

4. Alternative assignments or assessments: Providing alternative assignments or assessments that assess the individual's knowledge and skills in a format that is more accessible to them, such as oral presentations or projects, can accommodate their difficulties with reading and writing.

5. Preferential seating: Allowing individuals with dyslexia to choose their seating arrangement can reduce distractions and enhance focus during instruction.

6. Dyslexia-friendly environments: Creating dyslexia-friendly classrooms or workspaces by using dyslexia-friendly fonts, organizing materials, and providing clear visual aids can improve readability and overall learning experience.

It is important to note that accommodations should be tailored to the individual's specific needs and should be implemented in a way that promotes independence and self-advocacy. Regular communication and collaboration with the individual, their parents/guardians, teachers, and support professionals are crucial in identifying and implementing appropriate accommodations.

Interventions for Dyslexia

In addition to accommodations, interventions play a vital role in supporting individuals with dyslexia in developing their reading and writing skills. These interventions are designed to address the underlying difficulties associated with dyslexia and improve overall literacy abilities.

Here are some evidence-based interventions commonly used for individuals with dyslexia:

1. Orton-Gillingham Approach: This structured, multi-sensory approach focuses on teaching phonemic awareness, phonics, decoding, and spelling through explicit, systematic instruction. It emphasizes the use of visual, auditory, and kinesthetic modalities to reinforce learning.

2. Multi-sensory Structured Language Education (MSLE): MSLE is similar to the Orton-Gillingham Approach and is designed to teach reading, writing, and spelling in a structured, sequential, and cumulative manner. It integrates visual, auditory, and kinesthetic-tactile techniques to enhance learning.

3. Phonological Awareness Training: Phonological awareness refers to the ability to recognize and manipulate the sounds of language. Phonological awareness training interventions focus on developing skills such as rhyming, segmenting and blending sounds, and manipulating syllables, which are essential for reading and spelling.

4. Assistive Technology: Assistive technology, such as text-to-speech software, speech recognition programs, and word prediction tools, can support individuals with dyslexia in reading, writing, and organization tasks. These tools can help compensate for specific difficulties and promote independence.

5. Reading Intervention Programs: Various reading intervention programs, such as Wilson Reading System, Lindamood-Bell Programs, and Fast ForWord, provide structured and intensive instruction to improve decoding, fluency, comprehension, and spelling skills in individuals with dyslexia.

It is important to note that interventions should be tailored to the individual's specific needs and should be implemented by qualified professionals, such as special education teachers, reading specialists, or speech-language pathologists. A comprehensive evaluation and ongoing progress monitoring are important to ensure the effectiveness of interventions and to make any necessary adjustments.

Examples and Resources

Here are some examples of how accommodations and interventions can be implemented for individuals with dyslexia:

- Example 1: A student with dyslexia is provided with extended time on exams, access to assistive technology, and explicit instruction in phonological awareness and decoding skills through the Orton-Gillingham

Approach. These accommodations and interventions enable the student to effectively demonstrate their knowledge and improve their reading abilities.

- Example 2: An adult with dyslexia working in an office environment is provided with dyslexia-friendly materials, access to assistive technology, and a mentor/coach who provides guidance and support. These accommodations and interventions facilitate the individual's productivity, organization, and overall well-being in the workplace.

Resources for accommodations and interventions for individuals with dyslexia include:

- International Dyslexia Association (IDA): The IDA provides information, resources, and professional development opportunities related to dyslexia. Their website offers guidelines for accommodations, evidence-based interventions, and other support materials.

- Learning Ally: Learning Ally is a nonprofit organization that provides audiobooks, assistive technology, and learning solutions for individuals with dyslexia and other print disabilities. Their resources can support students in accessing reading materials and improving reading skills.

- National Center on Improving Literacy (NCIL): The NCIL offers evidence-based literacy resources, tools, and professional development opportunities for individuals with dyslexia. Their focus is on building literacy skills through assessment, intervention, and support.

Remember, implementing effective accommodations and interventions requires ongoing assessment, collaboration, and adjustment based on the individual's progress and needs. With the right support, individuals with dyslexia can overcome challenges and thrive in their educational and work settings.

The Role of Assistive Technology in Supporting Individuals with Dyslexia

Assistive technology plays a crucial role in supporting individuals with dyslexia by providing tools and resources that help them overcome challenges and enhance their learning experience. Dyslexia is a neurodevelopmental condition that affects reading, writing, and spelling skills. Individuals with dyslexia often struggle with decoding words, recognizing letters, and comprehending written text. However, with the help of assistive technology, these individuals can improve their reading and writing abilities, boost their confidence, and succeed academically.

1. Understanding Dyslexia and its Challenges

Dyslexia is characterized by difficulties in phonological processing, which is crucial for reading and writing. The brain processes written language differently in individuals with dyslexia, making it harder for them to decode words and associate sounds with letters. The challenges faced by individuals with dyslexia include:

a) Difficulty in reading fluently and accurately. b) Poor spelling and writing skills. c) Challenges in understanding and comprehending written text. d) Slow reading speed. e) Difficulty with word recognition and phonemic awareness.

2. Assistive Technology for Reading and Decoding

Assistive technology can provide targeted support for individuals with dyslexia to improve their reading and decoding skills. Some commonly used assistive technology tools for reading and decoding include:

a) Text-to-speech software: These tools convert written text into spoken words, allowing individuals with dyslexia to listen to the text while following along. This helps improve reading accuracy, fluency, and comprehension.

b) Optical character recognition (OCR) software: OCR software scans printed text and converts it into editable and readable digital text. Individuals with dyslexia can use this tool to access printed materials in a format that is easier for them to read.

c) Reading apps: There are various apps available that provide dyslexia-friendly features such as dyslexia-friendly fonts, background color customization, and text highlighting. These features can make reading more accessible and enjoyable for individuals with dyslexia.

3. Assistive Technology for Writing and Spelling

In addition to reading support, assistive technology also offers tools to assist individuals with dyslexia in improving their writing and spelling skills. Some commonly used assistive technology tools for writing and spelling include:

a) Word prediction software: These tools suggest words as individuals type, helping them with spelling and reducing the effort required to recall and write words correctly.

b) Speech-to-text software: Speech-to-text tools convert spoken language into written text, allowing individuals with dyslexia to express their ideas through speech and have them transcribed into written form. This can help overcome spelling challenges and improve writing skills.

c) Grammar and spell-check tools: Assistive technology tools with built-in grammar and spell-check features can assist individuals with dyslexia in identifying and correcting spelling and grammatical errors in their written work.

4. Additional Features and Considerations

When selecting and using assistive technology for dyslexia, it is essential to consider the individual needs and preferences of the user. Some additional features and considerations include:

a) Customizable settings: Assistive technology tools that allow users to customize features such as font size, color contrast, and reading speed can make reading and writing more comfortable and accessible for individuals with dyslexia.

b) Multi-sensory approaches: Some assistive technology tools incorporate multi-sensory approaches, such as highlighting words as they are read aloud or using images and diagrams to support comprehension. These approaches can enhance the learning experience for individuals with dyslexia.

c) Training and support: It is essential to provide training and support for individuals using assistive technology to ensure they understand how to effectively utilize the tools. Training can help individuals with dyslexia maximize the benefits of assistive technology and improve their overall learning outcomes.

5. Real-World Examples

Assistive technology for dyslexia has been widely adopted in educational settings to support students with dyslexia. For example, Kurzweil 3000, a popular text-to-speech software, has been successfully used in many schools to improve reading comprehension and writing skills for students with dyslexia. Similarly, speech-to-text software like Dragon Naturally Speaking has helped individuals with dyslexia overcome spelling challenges and improve their written communication.

6. Conclusion

Assistive technology plays a vital role in supporting individuals with dyslexia by providing tools and resources that address their specific challenges. These technologies enhance reading, writing, and spelling skills, empower individuals with dyslexia, and promote their academic success. By recognizing the importance of assistive technology in supporting individuals with dyslexia, we can create more inclusive learning and working environments that celebrate the neurodiversity of all individuals.

Intellectual Disabilities

Definitions and Criteria for Intellectual Disabilities

Intellectual disabilities are a complex and diverse range of conditions that affect an individual's intellectual functioning and adaptive behavior. These disabilities, also known as intellectual developmental disorders, are typically characterized by deficits in intelligence, learning, and problem-solving skills. In this section, we will explore the definitions and criteria used to diagnose intellectual disabilities, as well as the impact they have on individuals and their lives.

Definition of Intellectual Disabilities

The American Association on Intellectual and Developmental Disabilities (AAIDD) defines intellectual disabilities as a disability characterized by significant limitations both in intellectual functioning and in adaptive behavior as expressed in conceptual, social, and practical adaptive skills. These limitations manifest before the age of 18.

Intellectual functioning refers to a person's general mental capacity and includes reasoning, problem-solving, abstract thinking, and learning from experience. Adaptive behavior, on the other hand, encompasses the skills necessary for daily life and includes communication, self-care, social interactions, and functional academics.

It is important to note that the AAIDD emphasizes that the definition of intellectual disabilities is not solely based on an individual's IQ score. Instead, it takes into account the overall combination of intellectual functioning and adaptive behavior.

Diagnostic Criteria for Intellectual Disabilities

To diagnose an intellectual disability, professionals use a combination of clinical assessments, observations, and standardized tests. The diagnostic criteria vary slightly depending on the diagnostic systems used, such as the Diagnostic and Statistical Manual of Mental Disorders (DSM-5) and the International Classification of Diseases (ICD-10).

The DSM-5 criteria for intellectual disabilities include the following:

1. Deficits in intellectual functioning: The individual must have an IQ below 70-75, as measured by a standardized test of intelligence. However, it is essential to interpret these scores cautiously, considering cultural and linguistic factors that may influence test performance.

2. Deficits in adaptive functioning: The individual must demonstrate limitations in adaptive behavior, such as difficulties with communication, self-care, social skills, and problem-solving. These limitations should be evident in multiple environments, such as home, school, or work settings.

3. Onset during the developmental period: The deficits in intellectual functioning and adaptive behavior must have onset before the age of 18. However,

the diagnosis can be made later in life if the individual's functioning was affected earlier but not recognized.

4. Significant impact on daily life: The intellectual disabilities should result in significant limitations in functioning, leading to the need for ongoing support and assistance in daily activities.

It is crucial to consider the individual's cultural and linguistic background when making a diagnosis of intellectual disabilities. Differences in language, educational opportunities, and cultural norms may influence performance on standardized tests and adaptive behavior measures.

Levels of Intellectual Disability

Intellectual disabilities are classified into different levels based on the level of support needed by individuals. The DSM-5 identifies the following levels of intellectual disability:

1. Mild intellectual disability: Individuals with mild intellectual disabilities typically have an IQ range of 50-70 and may have minimal difficulties with academic and social skills. They can often achieve independence and self-sufficiency with appropriate support and accommodations.

2. Moderate intellectual disability: Individuals with moderate intellectual disabilities usually have an IQ range of 35-50. They may experience challenges with academic and social skills, requiring ongoing support and individualized instruction. With appropriate support, they can learn practical skills and achieve a reasonable level of independence.

3. Severe intellectual disability: Individuals with severe intellectual disabilities usually have an IQ range of 20-35. They have significant limitations in academic, social, and self-help skills and may require extensive supervision and support. They can benefit from structured environments and learning opportunities.

4. Profound intellectual disability: Individuals with profound intellectual disabilities typically have an IQ below 20. They have severe impairments in intellectual functioning, communication, and self-care skills. They require constant support and care throughout their lives.

Understanding Intellectual Disabilities within the Neurodiversity Framework

The concept of neurodiversity recognizes that neurological differences, including intellectual disabilities, are a natural and valuable part of human diversity. It emphasizes the importance of recognizing and accepting individuals with intellectual disabilities for who they are, rather than viewing them solely through a deficit-based lens.

Within the neurodiversity framework, individuals with intellectual disabilities are seen as having unique strengths and abilities that can contribute to society. It promotes a shift from a model of "normalization" or "fixing" individuals with intellectual disabilities to one of acceptance, inclusion, and accommodation.

It is essential for society to create an inclusive environment that respects the rights and dignity of individuals with intellectual disabilities. This includes providing equal opportunities for education, employment, and community participation, as well as ensuring access to appropriate supports and accommodations.

Conclusion

This section explored the definitions and criteria for diagnosing intellectual disabilities. We learned that intellectual disabilities are characterized by significant limitations in intellectual functioning and adaptive behavior. The diagnostic criteria emphasize the importance of assessing both of these areas and considering the impact on daily functioning. Additionally, the section discussed the levels of intellectual disability and the role of the neurodiversity framework in understanding and embracing individuals with intellectual disabilities.

Understanding and accepting the diversity of intellectual abilities in society is crucial for fostering a more inclusive and equitable world. By recognizing the strengths and unique perspectives of individuals with intellectual disabilities, we can strive to create a society that celebrates and supports their presence.

Understanding Intellectual Disabilities within the Neurodiversity Framework

In the context of neurodiversity, it is essential to have a comprehensive understanding of intellectual disabilities. Intellectual disabilities are a group of neurodevelopmental conditions characterized by significant limitations in intellectual functioning and adaptive behavior. These limitations manifest during the developmental period and affect an individual's ability to effectively participate in everyday activities.

Defining Intellectual Disabilities

Intellectual disabilities are typically diagnosed based on intelligence quotient (IQ) assessments and evaluations of adaptive behavior. Intelligence is measured using standardized tests that assess cognitive abilities across various domains, such as verbal comprehension, perceptual reasoning, working memory, and processing speed. Adaptive behavior refers to the skills necessary for independent functioning, including communication, self-care, social skills, and functional academics.

To receive a diagnosis of an intellectual disability, an individual's IQ score is usually below 70, and their adaptive behavior is significantly impaired. However, it is important to note that intelligence and adaptive behavior are not fixed measures but rather dynamic and influenced by various factors such as cultural background, education, and social experiences.

Neurodiversity Perspective on Intellectual Disabilities

Within the neurodiversity framework, intellectual disabilities are viewed as natural variations of human neurology rather than as disorders that need to be cured or normalized. This perspective emphasizes celebrating the unique strengths, abilities, and contributions of individuals with intellectual disabilities, while acknowledging the challenges they may face in a society designed primarily for neurotypical individuals.

The neurodiversity paradigm asserts that neurological differences, including intellectual disabilities, should be recognized and respected as part of the natural diversity of human cognition. It aims to shift the focus from a deficit-oriented mindset to a strengths-based approach, celebrating the diversity of cognitive abilities and fostering inclusive environments that accommodate and appreciate individuals with intellectual disabilities.

Enhancing Quality of Life and Inclusion

Inclusive practices are crucial in promoting the well-being and inclusion of individuals with intellectual disabilities. By providing appropriate supports and accommodations, it is possible to enhance their quality of life and enable active participation in various aspects of society, including education, employment, and community life.

Education plays a vital role in empowering individuals with intellectual disabilities. Inclusive educational settings that supplement traditional academic instruction with individualized supports enable students to develop their strengths, build relationships, and acquire functional skills. The principles of Universal Design for Learning (UDL) can guide educators in creating accessible and inclusive learning environments that address the diverse needs of students with intellectual disabilities.

Employment opportunities are critical for fostering the independence and self-determination of individuals with intellectual disabilities. By recognizing and harnessing the unique strengths and talents of these individuals, employers can create inclusive workplaces and benefit from diverse perspectives and skills. The provision of reasonable accommodations, such as job coaching, assistive technology, and modified work schedules, can significantly support the success of individuals with intellectual disabilities in the workplace.

Additionally, community support services, such as residential programs, recreational activities, and healthcare, contribute to the overall well-being of individuals with intellectual disabilities. These services should be designed with a person-centered approach, fostering self-advocacy, autonomy, and social inclusion.

Challenges and Solutions

Individuals with intellectual disabilities may face several challenges related to communication, social interaction, independent living, and self-regulation. These challenges can impact their ability to navigate everyday situations and participate fully in society. It is important to develop strategies and provide supports to address these challenges effectively.

Communication difficulties are common among individuals with intellectual disabilities. Augmentative and alternative communication (AAC) systems, including sign language, pictorial symbols, and speech-generating devices, can facilitate effective communication and enhance social interaction.

Social skills training programs and interventions can help individuals with intellectual disabilities develop social relationships, improve understanding of social cues, and enhance their overall social competence. Peer support networks and social groups provide opportunities for establishing connections and fostering a sense of belonging.

Independent living skills, such as personal hygiene, meal preparation, and household management, can be developed through structured training programs and supported living arrangements. These programs emphasize skill building, task analysis, and individualized support to promote independence and self-sufficiency.

Self-regulation and emotional well-being can be addressed through the provision of stress management techniques, mindfulness-based interventions, and access to mental health services. These approaches help individuals with intellectual disabilities develop coping mechanisms and enhance their emotional resilience.

It is crucial to involve and support families and caregivers in the journey of individuals with intellectual disabilities. Collaborative partnerships between families, educational institutions, service providers, and community organizations ensure a holistic approach to support individuals across different life domains.

Unconventional Approach: Embracing Neurodivergent Contributions

An unconventional yet vital aspect to consider within the neurodiversity framework is the recognition and celebration of the unique contributions individuals with intellectual disabilities bring to society. Their perspectives, creativity, and divergent thinking can lead to innovative problem-solving approaches and novel insights. By valuing and incorporating neurodivergent contributions, society can benefit from a more diverse and inclusive range of perspectives and experiences.

For example, some individuals with intellectual disabilities may excel in areas such as art, music, or manual skills. By providing platforms for their creative expression and recognizing their talents, society can embrace these valuable contributions and challenge traditional notions of intelligence and success.

In conclusion, understanding intellectual disabilities within the neurodiversity framework requires recognizing the unique strengths and challenges of individuals with intellectual disabilities. By adopting a strengths-based approach, promoting inclusive practices, and providing necessary supports, society can foster the well-being, inclusion, and contributions of individuals with intellectual disabilities. Embracing neurodiversity in all its forms enriches society and promotes a more inclusive and accepting world.

Enhancing Quality of Life and Inclusion for Individuals with Intellectual Disabilities

Enhancing the quality of life and promoting inclusion for individuals with intellectual disabilities is a fundamental aspect of the neurodiversity framework. It requires a multidimensional approach that encompasses various domains such as education, healthcare, social support, and community engagement. In this section, we will explore strategies and initiatives aimed at improving the well-being and integration of individuals with intellectual disabilities.

Understanding Intellectual Disabilities

Before delving into enhancing the quality of life for individuals with intellectual disabilities, it is crucial to have a thorough understanding of what intellectual disabilities entail. Intellectual disabilities, also known as intellectual developmental disorders, are characterized by significant limitations in intellectual functioning and adaptive behaviors. These limitations manifest during the developmental period and may impact an individual's ability to reason, problem-solve, and effectively respond to everyday challenges.

The diagnosis of intellectual disabilities is based on a comprehensive assessment that considers intellectual functioning, adaptive behaviors, and the age at which the limitations emerge. The severity of intellectual disabilities can vary significantly, ranging from mild to profound. It is important to note that individuals with intellectual disabilities have diverse strengths and abilities, and their potential for growth and development should not be underestimated.

Person-Centered Planning

Enhancing the quality of life for individuals with intellectual disabilities begins with person-centered planning. Person-centered planning is an approach that places the individual at the center of decision-making and empowers them to actively participate in setting goals and determining the strategies to achieve those goals. This approach recognizes the unique needs, preferences, and aspirations of each individual and promotes their autonomy and self-determination.

Person-centered planning involves a collaborative process that includes the individual, their family, caregivers, and a multidisciplinary team of professionals. It aims to identify the individual's strengths, interests, and support needs across various life domains, including education, employment, social relationships, and recreational activities. By actively involving the individual in the planning process, person-centered planning fosters a sense of ownership and control, thereby promoting self-advocacy and self-empowerment.

Individualized Supports and Services

Individualized supports and services play a pivotal role in enhancing the quality of life and promoting inclusion for individuals with intellectual disabilities. These supports are tailored to the unique needs and goals of each individual and aim to maximize their independence, self-sufficiency, and overall well-being. Some key areas of individualized supports and services include:

1. Education: Individualized education plans (IEPs) are developed to provide specialized educational supports and accommodations that address the specific learning needs of individuals with intellectual disabilities. This may include modifications in instructional methods, curriculum adaptations, assistive technologies, and specialized interventions aimed at fostering academic, social, and life skills.

2. Healthcare: Accessible healthcare services are essential for individuals with intellectual disabilities to maintain their physical and mental well-being. Healthcare providers should be trained in providing culturally competent care and have a thorough understanding of the unique healthcare needs and challenges faced by individuals with intellectual disabilities. In addition, appropriate screenings, assessments, and interventions should be in place to address any co-occurring medical conditions or mental health concerns.

3. Social Support: Building and maintaining social connections is crucial for individuals with intellectual disabilities. Social support networks, such as peer mentorship programs, community-based organizations, and support groups, can provide opportunities for individuals to engage in meaningful relationships, participate in community activities, and develop a sense of belonging and inclusion.

4. Vocational Rehabilitation: Promoting employment opportunities and vocational skills training is vital for individuals with intellectual disabilities to achieve economic independence and social integration. Vocational rehabilitation programs offer tailored supports, job coaching, and skill development initiatives that enable individuals to overcome barriers in the workplace, enhance their employability, and succeed in a variety of career paths.

5. Recreation and Leisure Activities: Inclusive recreational programs and leisure activities are essential for individuals with intellectual disabilities to engage in enjoyable activities, develop talents, and experience a sense of fulfillment. Community centers, adaptive sports programs, art workshops, and cultural events provide opportunities for individuals to explore and cultivate their interests, thereby enhancing their overall quality of life.

Community Engagement and Advocacy

Creating an inclusive and accepting community is integral to enhancing the quality of life and promoting the full inclusion of individuals with intellectual disabilities. Community engagement and advocacy initiatives can foster a culture of acceptance, equal opportunities, and respect. Some strategies for community engagement and advocacy include:

1. Awareness and Education: Awareness campaigns and educational initiatives can help dispel stereotypes, debunk misconceptions, and promote understanding and acceptance of individuals with intellectual disabilities. These initiatives can be targeted towards schools, workplaces, community organizations, and the general public.

2. Accessibility and Universal Design: Ensuring physical and environmental accessibility is vital for individuals with intellectual disabilities to fully participate

in community activities. This includes accessible transportation, public spaces, buildings, and information resources. Universal design principles should be incorporated to create inclusive environments that accommodate the diverse needs of all individuals.

3. Collaboration and Partnerships: Collaboration between various stakeholders, including individuals with intellectual disabilities, their families, community organizations, businesses, educational institutions, and government bodies, is crucial for driving systematic change and creating sustainable inclusive practices. By working together, these stakeholders can collectively address barriers and develop innovative solutions that promote inclusion and quality of life.

4. Advocacy and Policy Development: Advocacy efforts can help amplify the voices and rights of individuals with intellectual disabilities. Advocacy organizations and individuals can work towards influencing policy development, promoting legislative changes, and ensuring equitable access to resources and services. This advocacy can be carried out at local, national, and international levels.

5. Community-Based Support Services: Community-based support services, such as respite care, supported living arrangements, and day programs, provide individuals with intellectual disabilities and their families with the necessary assistance and resources to lead fulfilling lives. These services should be person-centered, culturally sensitive, and responsive to the unique needs and preferences of individuals and their families.

Case Study: Enhancing Inclusion in Education

Inclusive education is a critical aspect of enhancing the quality of life and promoting the inclusion of individuals with intellectual disabilities. Let's consider a case study to explore the practical implementation of inclusive education strategies.

Case Study: Emily is a 10-year-old girl with mild intellectual disability. She attends a mainstream elementary school and is included in a general education classroom with appropriate supports and accommodations. The school has implemented a range of inclusive strategies to foster Emily's learning and social engagement:

1. Differentiated Instruction: Emily's teachers use differentiated instruction techniques to address her individual learning needs and preferences. They provide visual aids, hands-on activities, and adjusted pacing to ensure Emily's active participation and understanding of the curriculum.

2. Individualized Supports: Emily has an individualized education plan (IEP) that outlines the specific accommodations and supports she requires. This includes

additional time for assignments, preferential seating, and the use of assistive technologies such as text-to-speech software to support her reading comprehension.

3. Peer Support and Collaboration: Emily is paired with a buddy student who provides peer support and acts as an inclusive role model. The buddy student assists Emily with class activities, fosters social interactions, and promotes positive relationships within the classroom.

4. Social Skills Training: Emily participates in social skills training sessions, facilitated by a school counselor or behavior specialist. These sessions focus on developing age-appropriate social interactions, communication skills, and conflict resolution strategies.

5. Parent and Teacher Collaboration: Emily's parents and teachers maintain regular communication to exchange information, monitor her progress, and address any concerns. This collaborative approach ensures a consistent support system for Emily and promotes a strong home-school partnership.

By implementing these inclusive education strategies, Emily's school provides a supportive and inclusive environment that allows her to thrive academically, develop social skills, and build meaningful relationships with her peers. These strategies not only enhance Emily's quality of life but also foster a culture of acceptance and inclusion within the entire school community.

Conclusion

Enhancing the quality of life and promoting inclusion for individuals with intellectual disabilities requires a comprehensive and person-centered approach. By implementing person-centered planning, individualized supports and services, community engagement and advocacy initiatives, and inclusive educational practices, we can create a society that celebrates the diversity and contributions of all individuals, regardless of their intellectual abilities. Recognizing and valuing the unique strengths and potentials of individuals with intellectual disabilities is a cornerstone of the neurodiversity framework, enabling them to lead fulfilling lives and fully participate in all aspects of society.

In the next section, we will explore the historical and present-day societal contexts that have shaped the perception and treatment of neurodivergent individuals, with a focus on the impact of discrimination and stigma.

Person-Centered Planning and Individualized Supports for Intellectual Disabilities

Person-centered planning and individualized supports play a crucial role in promoting the well-being and quality of life for individuals with intellectual disabilities. This approach focuses on tailoring support systems to meet the unique needs, preferences, and goals of each individual, empowering them to actively participate in decision-making processes and lead fulfilling lives. In this section, we will explore the principles of person-centered planning, the key components of an individualized support plan, and the importance of collaboration and advocacy in ensuring its successful implementation.

Principles of Person-Centered Planning

Person-centered planning is guided by a set of core principles that prioritize the rights, autonomy, and dignity of individuals with intellectual disabilities. These principles serve as a foundation for the development of individualized support plans and are essential in promoting self-determination and inclusion. Some key principles include:

- **Respect for Autonomy:** Recognizing and honoring the individual's right to make decisions about their own life and ensuring their active participation in the planning process.

- **Inclusion and Community Participation:** Promoting opportunities for individuals with intellectual disabilities to actively engage in their communities, access resources, and build social connections.

- **Collaboration and Partnerships:** Fostering collaborative relationships between individuals, their families, support professionals, and community members to develop comprehensive support plans that reflect the individual's needs and aspirations.

- **Strengths-Based Approach:** Identifying and building upon the individual's strengths, abilities, and interests to enhance their independence and overall well-being.

- **Flexibility and Adaptability:** Recognizing that individuals' needs and goals may change over time, and ensuring that support plans are flexible and responsive to these changes.

By adhering to these principles, person-centered planning promotes the empowerment and self-advocacy of individuals with intellectual disabilities, providing them with a sense of ownership and control over their own lives.

Components of an Individualized Support Plan

A comprehensive individualized support plan encompasses various components that are tailored to the unique needs and goals of the individual with an intellectual disability. These components include:

1. **Assessment of Support Needs:** Conducting a thorough assessment of the individual's strengths, challenges, preferences, and aspirations to inform the development of the support plan.

2. **Goal Setting:** Collaboratively determining short-term and long-term goals that reflect the individual's aspirations and are aligned with their support needs. Goals should be specific, measurable, attainable, relevant, and time-bound (SMART).

3. **Support Strategies and Interventions:** Identifying and implementing evidence-based strategies and interventions that promote the individual's independence, skill development, and overall well-being. This may include educational support, vocational training, therapy services, and community-based programs.

4. **Communication and Decision-Making Processes:** Ensuring that communication and decision-making processes are accessible and tailored to the individual's abilities. This may involve the use of visual supports, assistive technology, or alternative communication methods.

5. **Support Team and Collaboration:** Engaging a multidisciplinary team of professionals, including educators, therapists, support staff, and community members, in the planning and implementation of supports. Collaboration among team members is essential to ensure the continuity and effectiveness of the support plan.

6. **Monitoring and Evaluation:** Regularly reviewing and evaluating the effectiveness of the support plan in achieving the individual's goals. Adjustments and modifications should be made based on the individual's progress and evolving needs.

7. **Transitions and Life Planning:** Addressing transitions throughout the individual's life span, including transitions from school to adulthood, employment, and independent living. Planning for the individual's future, including health care and financial considerations, should also be included.

By incorporating these components into the support plan, individuals with intellectual disabilities can receive the necessary assistance and resources to navigate various aspects of their lives and achieve their full potential.

Collaboration and Advocacy

Person-centered planning and individualized supports rely on collaboration and advocacy to ensure their successful implementation. Collaboration involves the active involvement of the individual, their families, support professionals, and community members in decision-making processes and the development of the support plan. This collaborative approach facilitates a holistic understanding of the individual's needs and fosters shared responsibilities in supporting their well-being.

Advocacy plays a critical role in promoting and protecting the rights of individuals with intellectual disabilities. Advocates work to challenge societal barriers, raise awareness, and ensure that individuals have equal access to education, employment, healthcare, and community resources. Advocates also play a key role in influencing policy changes and promoting inclusive practices within various systems.

To effectively advocate for individuals with intellectual disabilities, it is important to:

- Understand and communicate the unique needs and challenges faced by individuals with intellectual disabilities.
- Collaborate with other advocates and organizations to amplify the voices and experiences of individuals with intellectual disabilities.
- Support self-advocacy efforts by empowering individuals to speak up for themselves and make informed decisions.
- Advocate for inclusive policies that promote access to quality education, employment, healthcare, and community resources.
- Build partnerships with stakeholders, such as policymakers, educators, employers, and healthcare providers, to drive systemic change and improve outcomes for individuals with intellectual disabilities.

By prioritizing collaboration and advocacy, person-centered planning can truly create an environment that respects and supports the aspirations and rights of individuals with intellectual disabilities.

Implementation Challenges and Strategies

While person-centered planning and individualized supports offer numerous benefits, their successful implementation can face various challenges. Some common challenges include:

- Limited resources: Availability of limited resources, including funding, specialized services, and trained professionals, can pose challenges in providing comprehensive support.
- Access barriers: Individuals with intellectual disabilities may face barriers in accessing services, supports, and community resources due to physical, social, or attitudinal barriers.
- Lack of coordination and communication between systems: Inadequate coordination and communication among different systems and service providers can lead to fragmented supports and information gaps.
- Attitudinal barriers: Prevailing stigmas, stereotypes, and biases can influence the attitudes and perceptions of professionals, families, and community members, impacting the quality of support provided.

To address these challenges, it is crucial to adopt strategies that promote effective implementation of person-centered planning and individualized supports. Some key strategies include:

- Advocating for increased funding and resources to expand access to services and supports.
- Promoting awareness and understanding of intellectual disabilities to challenge stigma and attitudinal barriers.
- Establishing effective communication and collaboration among service providers, families, and community members through regular meetings, shared documentation, and joint decision-making processes.

- Building capacity and providing professional development opportunities for support professionals to enhance their knowledge and skills in person-centered planning and evidence-based interventions.

- Encouraging partnerships between organizations, communities, and businesses to create inclusive environments and opportunities for individuals with intellectual disabilities.

By addressing these challenges and implementing effective strategies, person-centered planning and individualized supports can be successfully integrated into the lives of individuals with intellectual disabilities, promoting their independence, well-being, and inclusion.

Summary

Person-centered planning and individualized supports aim to empower individuals with intellectual disabilities, promoting their self-determination, inclusion, and overall well-being. It is guided by core principles that prioritize autonomy, collaboration, and the recognition of individuals' strengths and abilities. A comprehensive individualized support plan encompasses various components, including needs assessment, goal setting, strategies and interventions, communication and decision-making processes, support team collaboration, monitoring and evaluation, and life planning. Collaboration and advocacy are crucial in ensuring the successful implementation of person-centered planning, while addressing implementation challenges requires strategies such as resource allocation, reducing access barriers, improving coordination and communication, and challenging attitudinal barriers. Through the integration of person-centered planning and individualized supports, individuals with intellectual disabilities can lead meaningful and fulfilling lives, actively participating in their communities and achieving their goals.

Other Neurodevelopmental Conditions

Tourette Syndrome

Tourette Syndrome (TS) is a neurodevelopmental disorder characterized by repetitive and involuntary vocal and motor tics. These tics are sudden, rapid, and non-rhythmic movements or vocalizations that occur repeatedly. They can range from mild to severe and can vary in frequency and complexity.

The precise cause of Tourette Syndrome is currently unknown, but research suggests that it may result from a combination of genetic and environmental factors. It is believed that abnormalities in certain brain regions and neurotransmitter imbalances, particularly involving dopamine, may contribute to the development of TS.

Individuals with Tourette Syndrome often experience a range of challenges and difficulties. The presentation of symptoms can vary widely among individuals, making it a heterogeneous condition. Tics typically begin in childhood, around the age of 6 or 7, and may peak during the early teenage years. For some individuals, tics may decrease or even disappear in adulthood, while for others, they persist into adulthood.

In addition to the primary symptoms of tics, individuals with Tourette Syndrome may also experience associated co-occurring conditions. These can include attention-deficit/hyperactivity disorder (ADHD), obsessive-compulsive disorder (OCD), anxiety disorders, mood disorders, and learning difficulties. The presence of these comorbidities can further complicate and impact the daily lives of individuals with TS.

Managing Tourette Syndrome requires a comprehensive and multidimensional approach. Treatment options include behavioral therapy, medication, and support from healthcare professionals, educators, and family members. Behavioral therapy, such as cognitive-behavioral therapy (CBT) and habit reversal training, can help individuals manage and reduce the intensity of their tics.

Medication can also play a role in managing symptoms. Selective serotonin reuptake inhibitors (SSRIs) and antipsychotic medications are commonly prescribed to help reduce the frequency and severity of tics. However, it is essential to note that medications may have side effects, and their use should be carefully monitored and adjusted in collaboration with a healthcare professional.

Support from the educational system is crucial for individuals with Tourette Syndrome. Teachers and school staff can provide accommodations and modifications to support the learning needs of students with TS. These may include preferential seating, extra time for assignments and exams, and the provision of assistive technology. Creating a supportive and inclusive classroom environment can significantly impact the academic success and well-being of students with TS.

It is important to address misconceptions and stereotypes surrounding Tourette Syndrome. People with TS may face social stigma and misunderstanding due to their visible tics, which can often lead to isolation and discrimination. Educating the broader community about TS and promoting acceptance and inclusion is vital for creating a more inclusive society.

Despite the challenges that Tourette Syndrome presents, many individuals with TS have found success and fulfillment in various areas of life. Their unique perspectives and strengths can contribute to their personal growth and society as a whole. By recognizing and celebrating the neurodiversity within Tourette Syndrome, we foster a more inclusive world that values and appreciates the full spectrum of human differences.

Resources:

- Tourette Association of America: `https://tourette.org` - CDC - Tourette Syndrome: `https://www.cdc.gov/ncbddd/tourette` - National Institute of Neurological Disorders and Stroke - Tourette Syndrome Fact Sheet: `https://www.ninds.nih.gov/Disorders/Patient-Caregiver-Education/Fact-Sheets/Tourette-Syndrome-Fact-Sheet`

Developmental Coordination Disorder (DCD)

Developmental Coordination Disorder (DCD), also known as dyspraxia, is a neurodevelopmental condition that affects a person's ability to coordinate movements and perform skilled motor tasks. Individuals with DCD often have difficulties with tasks such as writing, tying shoelaces, and playing sports, which can significantly impact their daily functioning and quality of life.

Definition and Diagnosis of DCD

DCD is characterized by marked motor impairments that interfere with an individual's ability to perform age-appropriate activities. These difficulties are not due to a general medical condition (such as cerebral palsy or muscular dystrophy) and cannot be better explained by intellectual disabilities or visual impairments. The onset of DCD usually occurs in early childhood and may persist into adolescence and adulthood.

To diagnose DCD, professionals typically conduct a comprehensive assessment that includes a thorough medical history, observation of motor skills, and standardized motor assessments. The Diagnostic and Statistical Manual of Mental Disorders (DSM-5) provides diagnostic criteria for DCD, which include:

- The motor skills deficit significantly impacts daily activities and academic achievements.
- The motor skills difficulties are not solely explained by intellectual disabilities or visual impairments.

- The motor skills difficulties are not due to a neurological condition that affects movement (e.g., cerebral palsy) or a general medical condition (e.g., muscular dystrophy).
- The motor skills difficulties are not better accounted for by another mental disorder (e.g., autism spectrum disorder).

It is important to note that DCD is a diagnosis of exclusion, meaning that other possible causes of motor impairments must be ruled out before making a diagnosis.

Understanding the Motor Profile of Individuals with DCD

Individuals with DCD exhibit a range of motor difficulties that can vary in severity and presentation. Some common characteristics of DCD include:

- Poor coordination and balance: Individuals with DCD often struggle with activities that require coordination of different body parts, such as catching a ball or riding a bike. They may also have difficulty maintaining balance and may appear clumsy.
- Difficulty with fine motor skills: Tasks that involve precise movements, such as handwriting or using scissors, can be particularly challenging for individuals with DCD. They may struggle with tasks that require manual dexterity and may have difficulty with activities of daily living, such as tying shoelaces or using utensils.
- Awkward gait and posture: Some individuals with DCD may have an awkward walking style or challenges with maintaining an upright posture.
- Poor spatial awareness and organization: Individuals with DCD may have difficulty understanding and navigating through space. They may have challenges with spatial concepts, such as left and right, and struggle with tasks that involve spatial organization, such as packing a bag or organizing belongings.
- Difficulties with motor planning and execution: Planning and executing actions in a coordinated manner can be challenging for individuals with DCD. They may have trouble with tasks that require multiple steps and may struggle with sequencing movements.

It is important to understand that the motor difficulties experienced by individuals with DCD are not due to muscle weakness or lack of intelligence. Rather, they arise from difficulties in motor planning, coordination, and execution.

Strengths and Challenges of Individuals with DCD

While individuals with DCD face significant challenges in their motor abilities, it is essential to recognize and celebrate their strengths. Some strengths commonly associated with individuals with DCD include:

- Creativity and innovation: Many individuals with DCD have unique problem-solving skills and demonstrate creative approaches to tasks. They often think outside the box and come up with inventive solutions.
- Unique perspectives: The challenges faced by individuals with DCD can provide them with a different perspective on the world. They may have a heightened awareness of their environment and display a keen eye for detail.
- Resilience and determination: Despite the difficulties they face, individuals with DCD often demonstrate remarkable resilience and determination. They are motivated to overcome challenges and achieve their goals.
- Strong verbal and social skills: While fine and gross motor skills may be affected, individuals with DCD often excel in verbal and social domains. They may have excellent communication skills and interpersonal abilities.

However, individuals with DCD may also face various challenges, which can have a significant impact on their daily lives. These challenges may include:

- Academic difficulties: Difficulties with handwriting, organization, and fine motor skills can impact academic performance. Individuals with DCD may struggle with tasks such as note-taking, completing written assignments, and participating in activities that require precise motor skills.
- Reduced participation in physical activities: Coordination and motor difficulties can make it challenging for individuals with DCD to participate in sports and other physical activities. This can impact their overall physical health and well-being.
- Low self-esteem and social difficulties: The motor challenges experienced by individuals with DCD can lead to feelings of frustration, embarrassment, and reduced self-confidence. They may struggle with peer relationships and may feel excluded from activities that require physical coordination.

- High levels of anxiety and stress: The challenges associated with DCD can lead to heightened anxiety and stress. Individuals with DCD may experience anxiety related to their motor abilities and may worry about social interactions and performance in academic and work settings.

- Impact on daily living skills: Difficulties with fine motor skills can impact the ability to perform everyday tasks independently, such as getting dressed, tying shoelaces, and managing personal care activities.

Accommodations and Interventions for Individuals with DCD

Individuals with DCD can benefit from various accommodations and interventions to support their motor development and functional abilities. Some strategies and interventions that can be helpful include:

- Occupational therapy: Occupational therapy focuses on improving an individual's ability to perform daily activities and participate in meaningful occupations. Occupational therapists can provide individualized interventions to address specific motor difficulties and help individuals develop functional skills.

- Environmental modifications: Modifying the physical environment can help individuals with DCD navigate their surroundings more effectively. Simple adjustments, such as using visual cues, providing clear instructions, and organizing workspaces, can make tasks more manageable.

- Assistive technology: Various assistive technologies can support individuals with DCD in their daily activities. Tools such as specialized pens, ergonomic keyboards, and voice recognition software can help enhance writing skills and increase productivity.

- Movement and coordination programs: Participation in structured movement and coordination programs, such as yoga, dance, or martial arts, can improve motor planning and coordination. These programs can also promote body awareness, spatial skills, and self-confidence.

- Self-regulation strategies: Learning self-regulation strategies, such as deep breathing exercises and mindfulness techniques, can help individuals with DCD manage stress and anxiety related to motor difficulties.

It is crucial to adopt a strengths-based approach when supporting individuals with DCD. By focusing on their strengths and providing appropriate accommodations and interventions, we can help individuals with DCD thrive and achieve their full potential.

Real-Life Example: Sarah's Journey with DCD

Let's consider the case of Sarah, a 10-year-old girl diagnosed with DCD. Sarah often struggles with tasks that require fine motor skills, such as handwriting and tying shoelaces. These difficulties have caused her frustration and impacted her self-esteem, as she feels different from her peers. Despite her challenges, Sarah has shown a remarkable talent for storytelling and imaginative play. Through storytelling, Sarah finds an outlet where she can express herself freely and creatively.

Sarah's parents and teachers work together to support her. They provide Sarah with assistive technology, such as a voice-to-text software, to help her with writing tasks. Sarah's classroom environment is modified to accommodate her needs, with visual cues and clear instructions provided to support her organization and task completion. Additionally, Sarah participates in occupational therapy sessions, where she learns fine motor exercises and strategies to improve her handwriting.

With these accommodations and interventions in place, Sarah's confidence and overall well-being improve. She continues to excel in her storytelling abilities, and her unique perspective enriches her interactions with peers and teachers. Sarah's journey with DCD highlights the importance of recognizing and nurturing the strengths of individuals with DCD, while providing them with the right support and accommodations to overcome their challenges.

Resources for Individuals with DCD

For individuals with DCD and their families, accessing reliable resources and support is crucial. Here are some organizations and websites that provide information, advocacy, and support for individuals with DCD:

- Developmental Coordination Disorder (DCD) Canada: A Canadian organization that offers resources and support for individuals with DCD and their families. They provide information on diagnosis, interventions, and strategies to manage daily challenges.

- American Occupational Therapy Association (AOTA): A professional organization that provides resources and information on occupational therapy interventions for individuals with DCD. Their website offers tips and suggestions for home, school, and community environments.

- Dyspraxia Foundation: A UK-based organization that offers information, support, and resources for individuals with DCD and their families. They provide publications, online forums, and helpline services.
- National Center on Physical Activity and Disability (NCPAD): A resource center that focuses on promoting physical activity and inclusion for individuals with disabilities, including DCD. Their website offers guidance on adapting physical activities and finding inclusive sports programs.
- Understood: An online platform that provides resources, articles, and personal stories on developmental differences, including DCD. They offer practical tips for parents, educators, and individuals with DCD.

These resources can serve as valuable sources of information, encouragement, and support for individuals with DCD and their families. It is essential to consult with professionals, such as occupational therapists, to develop an individualized plan that addresses the specific needs and challenges of individuals with DCD.

Note: The content provided here is meant for informational purposes only and should not replace professional advice. If you suspect you or someone you know has DCD, it is important to consult with a qualified healthcare professional for a comprehensive assessment and diagnosis.

Specific Learning Disorders (SLD)

Specific Learning Disorders (SLD) are a group of neurodevelopmental disorders characterized by difficulties in acquiring and using academic skills, despite the absence of intellectual disabilities or other cognitive impairments. These disorders significantly impact an individual's ability to read, write, spell, or perform mathematical calculations. In this section, we will explore the definition, diagnosis, and interventions for specific learning disorders, focusing on dyslexia and dyscalculia.

Definition and Diagnosis of Dyslexia

Dyslexia is a specific learning disorder that primarily affects reading skills. Individuals with dyslexia struggle with accurate and fluent word recognition and have difficulty decoding written words. This can result in challenges with reading comprehension and slower reading rates compared to their peers. Dyslexia is not related to intelligence or vision problems, but rather stems from differences in the way the brain processes language.

Diagnosing dyslexia involves a comprehensive evaluation conducted by a qualified professional, such as a psychologist or educational specialist. The evaluation typically includes assessments of reading and language abilities, analysis of the individual's educational and developmental history, and consideration of any other factors that may be impacting their reading performance.

Understanding the Learning Profile of Individuals with Dyslexia

It is important to recognize that individuals with dyslexia often possess unique strengths and abilities. While they may struggle with certain aspects of reading, they may excel in other areas, such as creative thinking, problem-solving, or spatial reasoning. Dyslexic individuals often have strong visual and spatial skills, allowing them to think outside the box and approach problems from different perspectives.

The learning profile of individuals with dyslexia can vary greatly. Some may benefit from multisensory teaching methods that engage multiple senses (such as visual, auditory, and kinesthetic) to enhance reading skills. Others may require additional support in areas such as phonemic awareness, phonics instruction, or reading fluency.

Strengths and Challenges of Individuals with Dyslexia

Dyslexia presents unique challenges for individuals, but it is also important to acknowledge their strengths. Some common strengths of individuals with dyslexia include:

- Excellent problem-solving skills
- Strong creativity and imagination
- Ability to think in pictures and visualize concepts
- Strong spatial reasoning abilities
- Out-of-the-box thinking and innovative ideas

However, dyslexic individuals may encounter various challenges throughout their academic journey, such as:

- Difficulties with reading accuracy and fluency

- Poor spelling and writing skills
- Challenges in reading comprehension
- Limited vocabulary and word retrieval difficulties
- Frustration and anxiety related to reading tasks

Accommodations and Interventions for Individuals with Dyslexia

Accommodations and interventions play a crucial role in supporting individuals with dyslexia. These strategies aim to provide equal access to education, enhance learning opportunities, and promote academic success. Some common accommodations and interventions for individuals with dyslexia include:

- Assistive technology: Using tools like text-to-speech software, word prediction software, or audiobooks can assist dyslexic individuals in accessing written information more easily.
- Multisensory instruction: Incorporating visual, auditory, and kinesthetic elements in teaching can improve learning and retention for dyslexic learners.
- Orton-Gillingham approach: This structured, systematic, and multisensory teaching method focuses on teaching phonemic awareness, phonics, and decoding skills.
- Accommodations for exams: Providing extra time, reading assistance, or allowing the use of assistive technology during exams can give dyslexic students a fair chance to demonstrate their knowledge.
- Teacher support and understanding: Educators who are knowledgeable about dyslexia can create supportive and inclusive classroom environments by implementing appropriate teaching strategies and offering personalized guidance.

The Role of Assistive Technology in Supporting Individuals with Dyslexia

Assistive technology plays a critical role in supporting individuals with dyslexia, providing them with tools to overcome reading and writing difficulties. Here are some examples of assistive technology commonly used by dyslexic individuals:

- Text-to-speech software: This technology converts written text into spoken words, enabling dyslexic individuals to listen to digital text instead of reading it.

- Word prediction software: These programs offer suggestions and completion options while typing, helping dyslexic individuals with spelling and word retrieval.

- Audiobooks and e-readers: Accessible formats of books, such as audiobooks or electronic text with adjustable fonts and colors, allow dyslexic individuals to engage with written content more effectively.

- Speech recognition software: This technology converts spoken words into written text and can be useful for dyslexic individuals who struggle with writing or spelling.

While assistive technology can be highly beneficial, it is important to remember that it is not a substitute for effective instruction. Rather, assistive technology should be used as a tool to support individuals with dyslexia in their learning journey.

Conclusion

In this section, we explored the specific learning disorder of dyslexia. We discussed its definition, diagnosis, and understanding the learning profile of individuals with dyslexia. We also highlighted the strengths and challenges of dyslexic individuals and provided various accommodations, interventions, and assistive technology options to support their educational journey. By acknowledging the strengths and leveraging appropriate interventions, we can create inclusive educational environments that empower dyslexic individuals to thrive academically.

Nonverbal Learning Disabilities (NVLD)

Nonverbal Learning Disabilities (NVLD) is a neurodevelopmental condition that falls under the broader category of learning disabilities. While it shares some similarities with other learning disabilities, NVLD is characterized by a specific deficit in nonverbal processing skills. This section will explore the definition and diagnosis of NVLD, the challenges faced by individuals with NVLD, and strategies for providing support and accommodations.

Definition and Diagnosis of NVLD

NVLD is often described as a "hidden disability" because individuals with NVLD often have average to above-average verbal abilities, while their nonverbal skills are significantly impaired. These nonverbal skills include visual-spatial processing, fine motor coordination, and social skills.

The diagnosis of NVLD can be challenging due to its overlap with other conditions such as autism spectrum disorder and specific learning disorders. However, there are certain criteria that can aid in its diagnosis. These include a significant discrepancy between verbal and nonverbal abilities, an impaired ability to read nonverbal cues, poor visual-spatial skills, difficulties with executive functions, and challenges with social interaction and adaptive behavior.

Strengths and Challenges of Individuals with NVLD

Individuals with NVLD often exhibit strengths in verbal communication and memorization skills. They may have a vast vocabulary, excellent reading comprehension, and strong verbal expression. Additionally, they often possess well-developed verbal reasoning abilities, which can contribute to their academic success.

However, individuals with NVLD also face significant challenges. Their impaired nonverbal skills can manifest in difficulties with visual-spatial tasks, such as reading maps or understanding geometric concepts. They may struggle with fine motor skills, leading to challenges with handwriting or activities that require manual dexterity. Socially, individuals with NVLD may have difficulty interpreting facial expressions, body language, or social cues, which can impact their relationships and ability to navigate social situations effectively.

Accommodations and Interventions for NVLD

Providing appropriate accommodations and interventions is crucial for supporting individuals with NVLD in academic, social, and daily living settings. Some strategies that can be effective include:

- **Visual supports:** Visual aids, such as diagrams, charts, and graphic organizers, can help individuals with NVLD better understand and organize information.

- **Explicit instruction:** Breaking down tasks and providing step-by-step instructions can support individuals with NVLD in completing activities successfully.

- **Social skills training:** Teaching explicit social skills, such as recognizing emotions and understanding social cues, can help individuals with NVLD navigate social interactions more effectively.
- **Environmental modifications:** Creating an organized and predictable environment can support individuals with NVLD in managing their challenges with visual-spatial processing and executive functions.
- **Assistive technology:** Using assistive technology, such as speech-to-text software or graphic organizers, can help individuals with NVLD compensate for their difficulties with fine motor skills and nonverbal processing.

It is important to remember that each individual with NVLD is unique, and interventions should be tailored to their specific strengths and challenges. Collaborating with a multidisciplinary team, including educators, therapists, and parents, can help develop a comprehensive support plan.

Addressing Emotional and Psychological Well-being

Individuals with NVLD may experience challenges with their emotional and psychological well-being due to their difficulties with social interactions and the potential for academic struggles. It is crucial to address these aspects to promote their overall well-being. Some approaches include:

- **Building self-esteem:** Fostering a positive self-image and celebrating their strengths can help individuals with NVLD develop confidence in their abilities.
- **Emotional regulation:** Providing strategies for managing emotions and coping with stress can assist individuals with NVLD in navigating challenging situations.
- **Social support:** Facilitating opportunities for social connections with peers who are understanding and accepting can create a supportive network for individuals with NVLD.
- **Counseling and therapy:** Offering individual or group therapy sessions focused on self-acceptance, social skills, and emotional well-being can be beneficial for individuals with NVLD.

It is essential to validate the experiences of individuals with NVLD and create an inclusive and accepting environment that promotes their mental health.

Real-life Application: Developing Visual-Spatial Skills

Developing visual-spatial skills is crucial for individuals with NVLD. One way to enhance these skills is through practical exercises and activities. For example, engaging in puzzles, such as jigsaw puzzles or block-building games, can improve visual-spatial reasoning and problem-solving abilities. Similarly, activities that involve mapping, such as orienteering or geocaching, can help individuals with NVLD develop their navigation and spatial awareness skills.

These activities can be incorporated into educational settings or pursued as hobbies to provide individuals with NVLD opportunities to practice and enhance their visual-spatial skills in a practical and enjoyable manner.

Conclusion

Nonverbal Learning Disabilities (NVLD) present unique challenges for individuals, with impairments in nonverbal skills and social interactions. However, with understanding, appropriate accommodations, and support, individuals with NVLD can overcome these challenges and thrive academically, socially, and emotionally. Through a collaborative and inclusive approach, we can celebrate the diverse talents and strengths of individuals with NVLD, nurturing a more inclusive and accepting society that values neurodiversity.

The Social and Cultural Context of Neurodiversity

The History of Neurodivergent Communities

Neurodiversity in Ancient Cultures

Neurodiversity, the recognition and celebration of natural variations in human neurology, is a concept that has gained significant traction in recent years. However, the idea of embracing neurodivergent individuals is not a new one. When we look back at ancient cultures, we can find traces of neurodiversity acceptance and inclusion. In this section, we will explore some examples of how neurodiversity was understood and valued in ancient civilizations.

Egyptian Civilization

One of the earliest records of neurodivergent individuals being valued can be found in the ancient Egyptian civilization. In ancient Egypt, people with neurodivergent traits were believed to possess special gifts and abilities. They were often respected and revered, as their unique perspectives and talents were seen as valuable contributions to society.

For example, it is believed that some individuals with what we now understand as autism spectrum traits may have been selected as "healers" or "seers" in ancient Egyptian culture. These individuals were thought to have a heightened sensitivity and connection to the spiritual realm, making them ideal candidates for divination or healing practices.

Ancient Egyptian hieroglyphics also depict scenes that suggest the existence of neurodivergent individuals in their society. These depictions show people with distinctive behaviors or physical attributes, indicating that they were recognized as different but still included within the community.

Greek and Roman Civilizations

In ancient Greece and Rome, neurodivergent individuals were also regarded with respect and admiration. Philosophers and scholars such as Plato and Aristotle recognized the unique perspectives and abilities of individuals who diverged from the neurotypical norm.

One notable figure from ancient Greece is Pythagoras, famous for his contributions to mathematics. It is believed that Pythagoras may have had traits consistent with what we now understand as neurodivergence, such as an intense focus and obsession with mathematical patterns. His brilliance and unconventional methods of problem-solving have left a lasting impact on the field of mathematics.

In ancient Rome, individuals with neurodivergent traits were often celebrated for their artistic talents. Romans understood that the unconventional thinking and creativity of these individuals led to the creation of unique and groundbreaking works of art and literature. Examples include the poet Virgil and the playwright Terence, both of whom are believed to have had neurodivergent traits.

Indigenous Cultures

Many indigenous cultures around the world have a long history of valuing and embracing neurodivergent individuals. These cultures often recognized the unique strengths and gifts that neurodivergent individuals brought to the community.

For instance, among the Navajo people of North America, individuals with neurodivergent traits were believed to possess a deep spiritual connection and were often seen as important spiritual leaders or healers. Their ability to perceive the world in a different way was considered a valuable asset in guiding the community.

Similarly, some indigenous Australian cultures believe that certain neurodivergent individuals possess heightened sensory perceptions and spiritual insights. They are often regarded as "clever people" and play important roles in storytelling, healing practices, and decision-making within the community.

Understanding Neurodiversity in Ancient Cultures

The examples from ancient civilizations highlight the existence of neurodivergent individuals throughout history and provide evidence of their acceptance and inclusion in various societies. These cultures understood that neurodiversity is not a flaw or a defect but rather a natural variation within the human experience.

By recognizing and celebrating the unique perspectives and abilities of neurodivergent individuals, these ancient civilizations were able to tap into the rich diversity of human potential. They understood that neurodiversity brought forth new ideas, creative solutions to complex problems, and a different way of perceiving the world.

These historical examples challenge the notion that neurodivergent traits should be pathologized or suppressed in modern society. Instead, they serve as a reminder that embracing neurodiversity can lead to a more inclusive and innovative future.

Conclusion

The recognition and celebration of neurodiversity have deep roots in ancient civilizations. The examples from ancient Egypt, Greece, Rome, and indigenous cultures demonstrate that neurodivergent individuals have long been recognized for their unique abilities and contributions to society.

By studying these historical perspectives, we can learn valuable lessons about the importance of acceptance, inclusion, and the celebration of differences in a neurotypical world. Understanding neurodiversity in ancient cultures helps us challenge outdated stereotypes and cultivate a more inclusive and supportive environment for all individuals, regardless of their neurology.

In the next section, we will explore the influence of psychiatry and psychology on neurodivergent individuals, tracing the evolution of how neurodiversity has been understood and approached in modern history.

The Influence of Psychiatry and Psychology on Neurodivergent Individuals

Psychiatry and psychology play a significant role in understanding and addressing the needs of neurodivergent individuals. These disciplines contribute to the diagnosis, treatment, and management of various neurodevelopmental conditions.

Diagnosis of Neurodivergent Conditions

Psychiatry and psychology help in diagnosing neurodivergent conditions such as autism spectrum disorder (ASD), attention-deficit/hyperactivity disorder (ADHD), and dyslexia. These diagnoses are based on standardized criteria outlined in diagnostic manuals such as the Diagnostic and Statistical Manual of Mental Disorders (DSM-5). These diagnostic criteria help professionals determine the presence and severity of specific symptoms associated with neurodivergent conditions.

However, it is important to note that the diagnostic process can be complex and multidimensional. Clinicians need to consider multiple factors, including behavioral observations, developmental history, and input from parents or caregivers. Collaboration between psychiatrists, psychologists, and other professionals, such as speech-language pathologists or occupational therapists, is often necessary to obtain a comprehensive understanding of an individual's neurodivergent profile.

Treatment Approaches

Psychiatry and psychology offer a range of treatment approaches for neurodivergent individuals, aiming to enhance their quality of life and overall functioning. These approaches may include:

1. Behavioral interventions: Applied Behavior Analysis (ABA) is a widely used intervention for individuals with ASD. It focuses on promoting positive behaviors and reducing challenging behaviors through systematic applications of principles derived from the field of behavior analysis.

2. Cognitive-behavioral therapy (CBT): CBT is a well-established approach used to address symptoms associated with various neurodivergent conditions, such as anxiety and depression. It helps individuals identify and modify negative thoughts and behaviors, develop coping strategies, and improve overall mental well-being.

3. Social skills training: Many individuals with neurodivergent conditions struggle with social interactions. Psychologists and therapists use structured interventions to teach social skills, enhance communication abilities, and promote inclusion in social settings.

4. Educational interventions: Specialized educational programs and individualized supports are essential for neurodivergent individuals to thrive academically. These interventions may include personalized learning plans, accommodations, and assistive technologies to facilitate learning and optimize educational outcomes.

5. Medication management: In some cases, psychiatrists may prescribe medication to manage specific symptoms associated with neurodivergent conditions. Medication can help mitigate challenges related to attention, hyperactivity, impulsivity, anxiety, and mood.

It is important to note that treatment approaches should always be individualized, taking into account the unique strengths, challenges, and needs of each neurodivergent individual. Collaborative decision-making involving the individual, their families, and multidisciplinary teams is crucial to ensuring the best possible outcomes.

Challenges and Ethical Considerations

Psychiatry and psychology face several challenges when it comes to working with neurodivergent individuals. These challenges include:

1. Stigma and discrimination: Despite efforts to promote acceptance and inclusion, stigma surrounding neurodivergent conditions persists. Psychiatrists and psychologists must be vigilant in challenging societal stereotypes and misconceptions, advocating for the rights and dignity of neurodivergent individuals.

2. Bias in diagnosis and treatment: There is a risk of overdiagnosis or underdiagnosis of neurodivergent conditions, primarily due to subjective interpretations of symptoms and individual differences in presentation. Professionals must be aware of their biases and strive to apply diagnostic criteria accurately and objectively.

3. Accessibility and cultural relevance: The field of psychiatry and psychology needs to address issues of accessibility and cultural relevance in diagnosis and treatment. It is crucial to consider cultural variations in understanding neurodivergent traits and develop interventions that respect diverse cultural perspectives.

4. Informed consent and autonomy: When working with neurodivergent individuals, ensuring informed consent and respecting their autonomy is of utmost importance. Professionals must strive to involve individuals in decision-making processes and empower them to make choices that align with their values and preferences.

Psychiatry, Psychology, and the Neurodiversity Paradigm

The concept of neurodiversity challenges traditional understandings of neurodevelopmental conditions and emphasizes the uniqueness and value of neurodivergent individuals. Within the neurodiversity paradigm, psychiatry and psychology are evolving to adopt a more inclusive and strengths-based approach.

This shift involves moving away from pathologizing neurodivergent traits and focusing on promoting acceptance, self-determination, and accommodation. It also involves recognizing the importance of fostering a supportive and inclusive society that values the contributions and perspectives of neurodivergent individuals.

Psychiatrists and psychologists are increasingly engaging with neurodivergent communities, listening to their experiences, and integrating their insights into research, practice, and policy development. This collaborative approach helps ensure that the voices of neurodivergent individuals are heard and respected in the fields of psychiatry and psychology.

Overall, psychiatry and psychology have a significant influence on understanding and supporting neurodivergent individuals. By recognizing the strengths and challenges associated with neurodivergence, these disciplines contribute to the creation of a more inclusive and equitable society. Continued research, advocacy, and interdisciplinary collaboration are essential to further enhance the well-being and quality of life for neurodivergent individuals.

From Institutionalization to Deinstitutionalization

In this section, we will explore the transition from institutionalization to deinstitutionalization within the context of neurodivergent communities. We will examine the historical developments, the challenges faced, and the positive outcomes that have emerged from this shift in approach.

Historical Background

In the early 20th century, neurodivergent individuals, particularly those with intellectual disabilities or mental health conditions, were often subjected to institutionalization. This practice involved confining individuals in large, isolated facilities, away from mainstream society. The underlying belief was that such individuals required strict control and supervision for their own safety and the safety of others.

However, over time, concerns regarding the ethical implications and the quality of life for institutionalized individuals began to emerge. Reports of abuse and neglect within these institutions shed light on the harmful consequences of this approach. Activists, disability rights advocates, and progressive thinkers started questioning the effectiveness and humanity of institutionalization.

Shift in Approach

The deinstitutionalization movement gained momentum in the latter half of the 20th century. This movement aimed to shift away from the isolation and control of institutionalization towards community-based care and support for neurodivergent individuals. The goal was to create a more inclusive society that valued the rights and dignity of all its members.

Deinstitutionalization involved the closure of large-scale institutions and the development of community-based alternatives. It emphasized the importance of providing individuals with the necessary supports and services to live fulfilling lives in their communities. This approach recognized that neurodivergent individuals, like anyone else, have the right to autonomy, independence, and inclusion.

Challenges and Benefits of Deinstitutionalization

The process of deinstitutionalization was not without its challenges. One major hurdle was the need to establish robust support systems in the community to ensure the well-being and integration of individuals transitioning from institutions. The lack of resources and the limited availability of alternative services posed significant barriers to smooth implementation.

Furthermore, the societal stigma associated with neurodivergence presented challenges in terms of acceptance and inclusion. Stereotypes and misconceptions about neurodivergent individuals, perpetuated by years of institutionalization, hindered the community's readiness to embrace new approaches. Education and awareness campaigns became essential in challenging these perceptions and promoting acceptance.

However, the benefits of deinstitutionalization were significant. Neurodivergent individuals gained greater control over their lives and experienced increased opportunities for personal growth and development. They were no longer confined to institutional settings but instead had the chance to participate actively in their communities, contribute their unique strengths, and cultivate meaningful relationships.

Moreover, deinstitutionalization paved the way for more person-centered and individualized approaches to care and support. This shift acknowledged the diversity of needs and preferences among neurodivergent individuals, allowing for tailored interventions and services. It recognized that a "one size fits all" approach was inadequate and that each person's journey in the community would be unique.

Lessons Learned and Future Considerations

As we reflect on the transition from institutionalization to deinstitutionalization, several lessons emerge. First, we recognize the importance of challenging societal attitudes and promoting a culture of acceptance and inclusion. By debunking stereotypes and fostering understanding, we create conditions for neurodivergent individuals to flourish and contribute fully to society.

Second, we acknowledge the need for comprehensive support systems that address neurodivergent individuals' diverse needs. This includes accessible healthcare, inclusive education, employment opportunities, and community programs that promote self-development and well-being. Collaboration among various stakeholders, including individuals, families, advocacy organizations, and policymakers, is crucial for the successful ongoing implementation of community-based support.

Lastly, we must remain cautious of potential pitfalls and challenges that may arise in the deinstitutionalization process. Adequate funding, training, and resources must be allocated to ensure the sustainability and effectiveness of community-based support systems. Maintaining vigilant oversight and evaluation mechanisms can help address any emerging issues promptly.

In conclusion, the shift from institutionalization to deinstitutionalization represents a significant milestone in the journey towards creating an inclusive society for neurodivergent individuals. By embracing community-based care and support, we afford them the dignity, autonomy, and opportunities they deserve. However, this ongoing process requires continuous effort, collaboration, and a commitment to challenging existing norms and stereotypes.

The Emergence of Self-Advocacy and Neurodivergent Organizations

The emergence of self-advocacy and neurodivergent organizations marks a significant turning point in the history of neurodiversity. It is a testament to the strength, resilience, and determination of neurodivergent individuals to have their voices heard and their rights recognized. In this section, we will explore the origins and growth of self-advocacy and neurodivergent organizations, their impact on policy and society, and the importance of their continued advocacy efforts.

Historical Background

For many years, neurodivergent individuals were largely excluded from conversations about their own lives and futures. Their experiences and perspectives were often ignored or disregarded in favor of medical professionals and caregivers making decisions on their behalf. However, the tide began to turn as more and more neurodivergent individuals started speaking up for themselves, demanding recognition, and advocating for their own rights and well-being.

The roots of self-advocacy can be traced back to the disability rights movement of the 1960s and 1970s. This movement pushed for equal rights and opportunities for people with disabilities, including those with neurodivergent conditions. The movement challenged the prevailing medical model that pathologized and stigmatized neurodivergent individuals and instead emphasized the importance of self-determination and community inclusion.

Formation of Neurodivergent Organizations

As the disability rights movement gained momentum, neurodivergent individuals began organizing themselves into self-advocacy groups and organizations. These grassroots initiatives aimed to empower neurodivergent individuals, provide them with a space to share their experiences, and advocate for their rights.

One of the earliest and most influential neurodivergent organizations is the Autistic Self Advocacy Network (ASAN), founded in 2006 by a group of autistic individuals. ASAN has been at the forefront of advocating for the rights and inclusion of autistic individuals in various areas of life, including education, employment, and healthcare. They have been instrumental in shaping policies and legislation that promote neurodiversity and self-determination.

Other neurodivergent organizations, such as the ADHD Self-Advocacy Network and Dyslexic Advantage, have also emerged to champion the rights and needs of specific neurodivergent communities. These organizations provide resources, support, and advocacy for individuals with ADHD and dyslexia, respectively.

Impact on Policy and Society

Self-advocacy and neurodivergent organizations have had a profound impact on policy and society. By amplifying the voices and experiences of neurodivergent individuals, these organizations have challenged prevailing attitudes and norms, leading to positive changes in how neurodivergence is understood and accommodated.

One of the notable contributions of self-advocacy organizations is the push for person-centered planning and individualized support. They have advocated for a shift away from a one-size-fits-all approach to support and services, emphasizing the importance of tailoring interventions to meet the unique needs and strengths of neurodivergent individuals.

Neurodivergent organizations have also played a crucial role in influencing education policy and practice. By advocating for inclusive education, they have promoted the idea that neurodivergent students have the right to access quality education alongside their neurotypical peers. These efforts have led to the development of inclusive educational models and the implementation of accommodations and supports in schools.

Moreover, self-advocacy and neurodivergent organizations have challenged negative stereotypes and stigma surrounding neurodivergence. They have worked tirelessly to promote a positive and accurate understanding of neurodiversity, highlighting the strengths and contributions of neurodivergent individuals to society.

Continued Advocacy Efforts

While significant progress has been made, there is still much work to be done in terms of achieving full acceptance and inclusion for neurodivergent individuals. Self-advocacy and neurodivergent organizations continue to advocate for systemic changes in policy, legislation, and societal attitudes.

One of the ongoing advocacy efforts is promoting the rights of neurodivergent individuals in the workplace. Neurodivergent organizations strive to create neurodiversity-friendly workplaces that value and accommodate the unique strengths and challenges of neurodivergent employees. They push for inclusive hiring practices, reasonable accommodations, and supportive work environments.

Another crucial area of advocacy is mental health support for neurodivergent individuals. Self-advocacy organizations recognize the higher prevalence of mental health challenges among neurodivergent individuals and advocate for accessible and culturally competent mental health services. They also emphasize the importance of promoting well-being and self-care among neurodivergent individuals.

In their advocacy efforts, self-advocacy and neurodivergent organizations collaborate with policymakers, educators, employers, and the broader community. They work to dismantle barriers and create a more inclusive and accepting society that values and embraces neurodivergent individuals.

Conclusion

The emergence of self-advocacy and neurodivergent organizations has been a pivotal development in the neurodiversity movement. These organizations have given a voice to neurodivergent individuals, challenging societal norms and advocating for their rights and well-being. Through their efforts, they have shaped policy, influenced education and employment practices, and fought against stigma and discrimination. The work of self-advocacy and neurodivergent organizations must continue as we strive for a truly inclusive and accepting society that celebrates the diversity of the human experience.

Discrimination and Stigma

The Medicalization and Pathologization of Neurodivergent Identities

In our society, the medicalization and pathologization of neurodivergent identities have had a profound impact on how we perceive and understand individuals who are neurodivergent. This section explores the historical and cultural aspects of this phenomenon, highlighting its implications for neurodivergent individuals.

The Historical Context

The medicalization of neurodivergent identities can be traced back to the early history of psychiatry and psychology. In the late 19th and early 20th centuries, medical professionals began to categorize and label various forms of human behavior and cognition that deviated from the norm. This categorization served as a way to understand and explain differences in human functioning, but it also led to the pathologization of neurodivergent individuals.

One of the earliest examples of pathologizing neurodivergent individuals is the classification of autism. Leo Kanner, a psychiatrist, first described autism in the 1940s as a psychiatric disorder. This categorization positioned autism as a deviation from the norm, reinforcing the notion that neurodivergent traits were inherently pathological and in need of treatment.

The Impact on Neurodivergent Individuals

The medicalization and pathologization of neurodivergent identities have had significant consequences for individuals who identify as neurodivergent. It has created a culture that sees neurodivergence as a problem to be fixed rather than a natural variation of human diversity. This perspective can lead to feelings of shame, self-doubt, and low self-esteem among neurodivergent individuals.

Moreover, the pathologization of neurodivergent identities often results in a focus on deficits and challenges rather than recognizing and celebrating the strengths and unique perspectives that neurodivergent individuals bring. By framing neurodivergence solely as a medical condition, the broader societal contributions and potential of neurodivergent individuals are often overlooked.

The Role of Diagnosis and Labels

Diagnoses and labels play a critical role in the medicalization and pathologization of neurodivergent identities. While a diagnosis can provide a sense of validation and access to support services, it can also reinforce the perception that neurodivergent individuals are fundamentally different from the neurotypical population.

Labels, such as "autistic" or "ADHD," can carry stigma and contribute to the marginalization of neurodivergent individuals. They can create stereotypes and misconceptions that perpetuate discrimination and prejudice. It is important to recognize that neurodivergent individuals are not defined solely by their diagnoses or labels but possess a range of strengths, abilities, and unique perspectives.

Moving Towards a Neurodiversity Paradigm

To challenge the medicalization and pathologization of neurodivergent identities, there has been a growing movement towards a neurodiversity paradigm. The neurodiversity paradigm recognizes neurodivergence as a natural and valuable form of human diversity rather than a disorder or deficit.

This paradigm acknowledges that neurodivergent individuals have unique strengths, perspectives, and ways of experiencing the world. By embracing neurodiversity, we can foster a society that celebrates and includes neurodivergent individuals, promoting acceptance, and understanding.

Shifting Perspectives

Shifting from a medical model to a neurodiversity paradigm requires a reevaluation of societal attitudes, policies, and practices. This shift involves:

1. Challenging harmful stereotypes and misconceptions about neurodivergent individuals through education and awareness campaigns. 2. Promoting inclusive approaches in education, employment, and healthcare that accommodate and support neurodivergent individuals. 3. Encouraging the participation of neurodivergent individuals in decision-making processes and ensuring their voices are heard and valued. 4. Advocating for policies and legislation that protect the rights and well-being of neurodivergent individuals. 5. Fostering a culture of acceptance, empathy, and diversity that values neurodivergent traits.

Creating an Inclusive Society

Creating an inclusive society that respects and celebrates neurodivergent identities benefits not only neurodivergent individuals but society as a whole. By embracing neurodiversity, we tap into a diverse pool of talents, skills, and perspectives that can drive innovation, creativity, and social progress.

It is essential to recognize that neurodivergent individuals have unique contributions to make and should be given equal opportunities to thrive. By moving away from medicalization and pathologization, we can create a society that values and supports neurodivergent individuals, fostering a more inclusive and compassionate world for everyone.

Key Takeaways: - The medicalization and pathologization of neurodivergent identities stem from historical categorization efforts in psychiatry and psychology. - The impact of this phenomenon on neurodivergent individuals includes stigma, shame, and overlooking their strengths and perspectives. - Diagnoses and labels play a role in reinforcing the medicalization and pathologization of neurodivergent identities. - Shifting towards a neurodiversity paradigm involves challenging stereotypes, supporting inclusion, and valuing the contributions of neurodivergent individuals. - Embracing neurodiversity benefits both neurodivergent individuals and society as a whole, fostering innovation and social progress.

Stereotypes and Misconceptions about Neurodivergent Individuals

Neurodivergent individuals, who have neurological conditions that differ from the neurotypical population, often face stereotypes and misconceptions that can lead to misunderstandings and discrimination. It is important to recognize and debunk these stereotypes in order to promote understanding and acceptance of neurodiversity. In this section, we will explore some common stereotypes and misconceptions about neurodivergent individuals and provide evidence and examples to challenge these beliefs.

One common stereotype is that neurodivergent individuals are "broken" or "abnormal." This perspective arises from a neurotypical bias, where the neurotypical population is considered the norm. However, it is crucial to recognize that neurodivergent individuals have different neurological wiring, not defective or lesser brains. Neurodiversity emphasizes that neurological differences are a natural variation of the human experience. By understanding and appreciating these differences, we can move away from viewing neurodivergent individuals as abnormal and instead celebrate their unique strengths and perspectives.

Another stereotype is that neurodivergent individuals lack empathy or social skills. This misconception stems from the fact that neurodivergent individuals may have difficulty with social communication or expressing emotions in conventional ways. However, research has shown that empathy exists in neurodivergent individuals, albeit it may be expressed differently. For example, individuals with autism spectrum disorder (ASD) may experience hyper-empathy, where they feel others' emotions intensely. By recognizing and appreciating these different ways of experiencing empathy, we can challenge the stereotype that neurodivergent individuals lack empathy.

There is also a misconception that neurodivergent individuals are intellectually impaired or intellectually superior. While it is true that some neurodivergent individuals may have intellectual disabilities or exceptional abilities in specific domains, it is important to remember that neurodiversity encompasses a wide range of conditions and abilities. For example, individuals with dyslexia may struggle with reading and writing but excel in areas such as creativity or problem-solving. By acknowledging and valuing the diverse strengths and abilities of neurodivergent individuals, we can challenge the stereotype that neurodiversity is solely defined by intellectual capacity.

Another stereotype is that neurodivergent individuals are a burden or require constant care. This stereotype fails to recognize the potential of neurodivergent individuals and perpetuates a narrative of dependency. In reality, many neurodivergent individuals are capable of leading independent and fulfilling lives with the right support and accommodations. By providing inclusive environments and removing barriers, society can empower neurodivergent individuals to thrive and contribute their unique talents to various domains, including education, employment, and the arts.

These stereotypes and misconceptions not only create barriers for neurodivergent individuals but also limit society's understanding of different perspectives and talents. To overcome these stereotypes, it is important to promote accurate information, challenge biases, and highlight positive narratives of neurodivergent individuals. By embracing neurodiversity and valuing the contributions of all individuals, we can foster a more inclusive and accepting society.

In conclusion, stereotypes and misconceptions about neurodivergent individuals often arise from a neurotypical bias and lack of understanding. By challenging these stereotypes and promoting accurate information, we can foster a more inclusive and accepting society. Neurodivergent individuals have diverse strengths and perspectives that should be celebrated. It is crucial to recognize their unique talents and contributions in various domains such as education, employment, and the arts. By embracing neurodiversity, we can create a world that celebrates differences and fosters the full potential of all individuals.

Ableism and Discrimination in Education

Ableism refers to the discriminatory attitudes and practices that devalue and marginalize individuals with disabilities, including neurodivergent individuals, in various areas of life, including education. This section explores the presence of ableism within educational settings and its impact on neurodivergent students. We will also discuss strategies for addressing and combatting ableism, promoting inclusivity, and creating an environment that supports the diverse needs of all students.

Understanding Ableism

Ableism is rooted in the concept of able-bodiedness, which assumes that people with disabilities are inferior to those without disabilities. In an educational context, ableism manifests in different ways, such as:

1. Segregation and Exclusion: Historically, students with disabilities were often segregated from their neurotypical peers and denied access to mainstream educational opportunities. This practice prevented neurodivergent students from receiving quality education and hindered their social integration.
2. Lower Expectations: Ableism can result in the underestimation of neurodivergent students' abilities and potential. Teachers, influenced by ableist perspectives, may have lower expectations of neurodivergent students, leading to limited opportunities for academic achievement and personal growth.
3. Lack of Accessibility: Educational environments that do not provide suitable accommodations and support for neurodivergent students reinforce ableist norms. Inaccessible classrooms, materials, and teaching methods can create barriers to learning and participation.
4. Negative Stereotyping: Neurodivergent students often face negative stereotypes that perpetuate ableism. These stereotypes portray them as deficient, disruptive, or incapable of reaching their full potential. These biased beliefs can lead to social exclusion, bullying, and lowered self-esteem.

Impacts of Ableism on Neurodivergent Students

Ableism in education can have significant negative effects on the well-being and academic success of neurodivergent students. Some common impacts include:

1. Academic Disadvantage: Without appropriate accommodations, neurodivergent students may struggle to access and engage with the curriculum. This can lead to lower academic achievement, limited educational opportunities, and increased dropout rates.

2. Social Isolation: Ableism can create a hostile environment for neurodivergent students, making it more challenging for them to form supportive relationships with peers. The resulting social isolation can lead to feelings of loneliness, depression, and anxiety.

3. Low Self-esteem: Constant exposure to ableist attitudes and practices can erode neurodivergent students' confidence in their abilities and worth. They may internalize ableist messages and develop negative self-perceptions, hindering their self-advocacy and overall well-being.

4. Barriers to Inclusion: Ableism can hinder the inclusion of neurodivergent students in extracurricular activities, school clubs, and other social events. This exclusion perpetuates the notion that neurodivergent individuals are somehow less deserving of participating fully in the school community.

Addressing Ableism and Promoting Inclusion

Creating an inclusive and equitable educational environment requires addressing ableism head-on and implementing strategies to promote the full participation and success of neurodivergent students. Here are some approaches that can help combat ableism in education:

1. Disability Awareness and Sensitivity Training: Educators should undergo training to raise awareness about ableism, dismantle stereotypes, and develop more inclusive practices. This training should include understanding different types of neurodivergence, embracing neurodivergent strengths, and learning effective teaching strategies for diverse learners.

2. Universal Design for Learning (UDL): UDL emphasizes the importance of designing lessons and educational materials that are accessible to all students, regardless of their learning differences. By incorporating multiple means of representation, engagement, and expression, UDL promotes inclusivity and allows neurodivergent students to fully engage in the learning process.

3. Individualized Education Plans (IEPs) and Accommodations: Individualized plans, such as IEPs, can ensure that neurodivergent students' unique needs are addressed. These plans should outline specific accommodations, modifications, and support services that allow students to access the curriculum and participate fully in the educational experience.

4. Anti-bullying and Inclusion Programs: Schools should implement programs that explicitly address and prevent bullying and discrimination. These programs can foster a culture of respect, empathy, and acceptance, creating a safe and inclusive learning environment for all students.

5. Peer Support and Mentoring Programs: Establishing peer support networks and mentoring programs can enhance the social inclusion of neurodivergent students. Connecting them with neurotypical peers who can offer friendship, guidance, and understanding can help break down barriers and facilitate positive social interactions.

6. Family and Community Engagement: Collaboration between families, educators, and the wider community is crucial for supporting neurodivergent students. By involving families in decision-making processes and ensuring their cultural and linguistic needs are met, schools can create a more inclusive and supportive environment.

7. Curriculum Diversification: Including diverse perspectives and experiences in the curriculum can counter ableism and promote understanding and acceptance. Incorporating literature, history, and art that represent diverse neurodivergent voices can challenge stereotypes and foster empathy.

Real-World Example: Neurodivergent Inclusion in a High School

Let's consider a real-world example of neurodivergent inclusion in a high school. The school has implemented a dedicated inclusion program that values the strengths and diverse needs of neurodivergent students. Some of the strategies they employ include:

1. Inclusive Learning Environment: The school provides a physically accessible and sensory-friendly environment, including quiet spaces for students to retreat to if needed. Classroom designs prioritize flexibility and minimize distractions.

2. Assistive Technology: The school invests in assistive technology devices and software to support the individual learning needs of neurodivergent students. These tools help with note-taking, organization, and reading comprehension, enabling students to access the curriculum more effectively.

3. Inclusive Pedagogies: Teachers use a variety of instructional methods, incorporating visual aids, hands-on activities, and alternative assessment formats to meet different learning styles. They also give students opportunities for group work, fostering collaboration and social skills development.

4. Peer Mentoring: The school pairs neurodivergent students with neurotypical peers who act as mentors, providing academic and emotional support. Mentors receive training on neurodiversity, making them effective allies and advocates for inclusion.

5. Awareness Campaigns: The school conducts regular awareness campaigns to challenge ableism and promote acceptance and empathy for neurodivergent

individuals. These campaigns involve classroom discussions, workshops, guest speakers, and school-wide events.

Through these strategies, the high school creates an environment where neurodivergent students feel supported, respected, and valued. As a result, these students experience improved academic outcomes, increased self-confidence, and stronger social connections.

Conclusion

Addressing ableism and promoting inclusion in education is crucial for creating learning environments that support the diverse needs of neurodivergent students. By raising awareness, providing accommodations, fostering inclusive pedagogies, and establishing support networks, educators can challenge ableist norms, combat discrimination, and empower neurodivergent students to reach their full potential. Ultimately, promoting neurodiversity in education benefits not only neurodivergent individuals but also the entire school community by fostering understanding, empathy, and a celebration of differences.

Neurodiversity and Employment Discrimination

Neurodiversity is a concept that promotes the idea that neurological differences, such as those associated with autism, ADHD, dyslexia, and other neurodevelopmental conditions, are natural variations of the human brain rather than deficits or disorders. It emphasizes the recognition and acceptance of these differences in the neurotypical world. However, despite the growing awareness and acceptance of neurodiversity, individuals with neurodivergent traits still face significant challenges in the workplace, including discrimination.

1. **Understanding Employment Discrimination:** Employment discrimination refers to the unfair treatment of an individual or a group of individuals based on their neurodivergent traits. This discrimination can occur at various stages, including the hiring process, promotions, job assignments, and in the general work environment. Neurodivergent individuals may face barriers due to biases, stereotypes, and lack of understanding or accommodation for their unique strengths and challenges.

2. **Legal Framework:** Legislation has been enacted in many countries to protect individuals with disabilities, including those with neurodivergent conditions, from employment discrimination. For example, in the United States, the Americans with Disabilities Act (ADA) prohibits discrimination against individuals with disabilities in all aspects of employment. Similarly, the United Kingdom has the Equality Act, which protects individuals with disabilities from discrimination in employment and

other areas of life. It is important for employers to be aware of these laws and their obligations to provide a fair and inclusive work environment.

3. **Types of Employment Discrimination:** Neurodivergent individuals may experience various types of discrimination in the workplace, including:

3.1 *Prejudice and Bias:* Neurodivergent individuals often face prejudice and bias based on stereotypes and misconceptions about their abilities and limitations. Employers may hold biased attitudes that can impact hiring decisions, job assignments, and career advancement opportunities.

3.2 *Lack of Accommodations:* Accommodations are necessary to level the playing field for neurodivergent individuals in the workplace. However, employers may fail to provide appropriate accommodations, such as flexible work schedules, task modifications, sensory support, or assistive technologies, thus hindering their ability to perform optimally.

3.3 *Isolation and Exclusion:* Neurodivergent individuals may feel isolated or excluded in the workplace due to difficulties with social interactions and communication. A lack of understanding from colleagues or supervisors can contribute to a hostile work environment, limiting their opportunities for growth and success.

3.4 *Bullying and Harassment:* The unique characteristics of neurodivergent individuals may make them more vulnerable to bullying and harassment. Employers need to address such behaviors and promote a safe and inclusive work environment.

4. **Addressing Employment Discrimination:** Overcoming employment discrimination requires proactive efforts from both employers and neurodivergent individuals. Here are some strategies:

4.1 *Awareness and Education:* Increase awareness and understanding of neurodiversity among employers and employees. Promote training sessions to challenge biases, debunk misconceptions, and foster empathy and inclusivity.

4.2 *Accommodations and Support:* Employers should implement reasonable accommodations to enable neurodivergent individuals to thrive in the workplace. This may include providing sensory supports, flexible work arrangements, clear communication channels, and assistive technologies tailored to individual needs.

4.3 *Inclusive Hiring Practices:* Employers can adopt inclusive hiring practices, such as job advertisements that emphasize diversity and inclusion, ensuring unbiased interview processes, and actively recruiting neurodivergent individuals.

4.4 *Mentoring and Support Programs:* Establish mentoring programs that pair neurodivergent individuals with experienced employees who can provide guidance and support. Encourage the formation of employee resource groups to create a supportive community within the organization.

4.5 *Promoting Workplace Culture:* Foster a culture of acceptance, understanding, and respect for neurodiversity. Encourage open communication and provide platforms for neurodivergent individuals to share their experiences and contribute to decision-making processes.

5. **Success Stories and Best Practices:** Many organizations have successfully embraced neurodiversity in the workplace by implementing inclusive policies and practices. For example, SAP, a multinational software corporation, launched the Autism at Work program, which focuses on hiring and nurturing the talents of individuals on the autism spectrum. Such initiatives not only benefit the neurodivergent employees but also contribute to increased innovation and productivity within the organization.

6. **Resources and Support:** Neurodivergent individuals and employers can seek support from various resources and organizations that specialize in neurodiversity and employment. These resources provide guidance on accommodations, legal rights, disclosure strategies, and best practices for fostering an inclusive work environment.

Overall, addressing employment discrimination against neurodivergent individuals requires a comprehensive approach that includes legal protections, awareness, education, accommodations, and a supportive workplace culture. Creating an environment that values and embraces neurodiversity not only benefits individuals with neurodivergent traits but also promotes innovation, diversity, and inclusivity within organizations.

Intersectionality and Neurodivergent Identity

In the context of neurodiversity, intersectionality refers to the idea that individuals with neurodivergent identities may also belong to various marginalized social groups based on factors such as race, gender, sexual orientation, socioeconomic status, and disability. Intersectionality recognizes that these different aspects of an individual's identity intersect to shape their experiences, challenges, and opportunities.

Neurodivergent individuals who also belong to marginalized social groups often face compounded challenges and forms of discrimination. For example, a neurodivergent person of color may experience both racism and ableism, which can further marginalize them in society. Similarly, a neurodivergent person who identifies as LGBTQ+ may face discrimination based on their sexual orientation, gender identity, and neurodivergent status.

Understanding the intersectionality of neurodivergent identity is crucial for several reasons. First, it helps challenge the notion of a monolithic "neurodivergent experience" and recognizes the diversity within the neurodivergent community.

Different individuals with neurodivergent identities experience their neurodivergence in unique ways, shaped by their intersecting social identities and cultural contexts.

Second, intersectionality highlights the importance of addressing multiple forms of discrimination and marginalization. By acknowledging the intersectional experiences of neurodivergent individuals, we can work towards creating inclusive and equitable spaces that address not only neurodivergence but also other social inequalities.

Third, intersectionality emphasizes the need for inclusive research, advocacy, and policies. It calls for a more comprehensive understanding of the complex and interconnected challenges faced by neurodivergent individuals. For instance, research on the experiences of neurodivergent individuals should consider the impact of intersectional identities to provide a more nuanced picture.

Moreover, intersectionality encourages collaboration and solidarity across different social justice movements. Recognizing the shared experiences of individuals facing multiple forms of discrimination can lead to collective action and advocacy for systemic change. It allows for the development of supportive networks, resources, and initiatives that address the intersectional needs of neurodivergent individuals.

To illustrate the concept of intersectionality in the neurodiversity context, let's consider the case of a neurodivergent woman from a low-income background. She may face barriers in accessing proper diagnosis, support, and accommodation due to financial constraints. Additionally, she might encounter gender-based discrimination and stereotypes that intersect with the ableism she experiences as a neurodivergent individual. Understanding the intersectionality of her identity helps us recognize and address the multiple layers of challenges she faces, advocating for targeted policies that consider her unique circumstances.

In conclusion, intersectionality is a vital framework in understanding the experiences of neurodivergent individuals. It emphasizes the need to consider the intersection of neurodivergent identity with other social identities and challenges faced by marginalized groups. By adopting an intersectional lens, we can work towards a more inclusive society that recognizes and supports the diverse experiences of neurodivergent individuals.

Cultural Perspectives on Neurodiversity

Cultural Variations in Understanding Neurodivergent Traits

Culture plays a significant role in shaping our understanding and perceptions of neurodivergent traits. Different cultural contexts can influence the way individuals and society perceive and respond to neurodiversity. In this section, we will explore the cultural variations in understanding neurodivergent traits and how they shape attitudes, beliefs, and practices related to neurodiversity.

Cultural perspectives on neurodivergent traits

Neurodivergent traits, such as those associated with autism, ADHD, dyslexia, and intellectual disabilities, are not universally understood in the same way across cultures. Cultural perspectives shape how neurodivergent behavior is perceived, diagnosed, and treated.

One example of cultural variation is in the understanding of social communication differences. In some cultures, certain behaviors associated with autism, like avoiding eye contact or not engaging in small talk, may be considered normal or even desirable. These cultures may value personal space and place less emphasis on direct eye contact during conversation.

Another example is the understanding of attention and hyperactivity. In Western cultures, ADHD is often defined by symptoms such as impulsivity, excessive talking, and difficulty maintaining focus. However, in other cultures, these behaviors may be seen as normal variations in behavior or even positive attributes, such as being lively, outgoing, or expressive.

Cultural variations also exist in the concept of dyslexia. In some cultures, difficulties with reading and writing may be attributed to language differences or learning styles rather than being labeled as a specific learning disorder. The perception of dyslexia can significantly influence the type and level of support provided to individuals with dyslexia within educational settings.

Cultural barriers to acceptance and inclusion

Although some cultures may have more accepting and inclusive attitudes towards neurodivergent traits, many cultures still face significant barriers to acceptance and inclusion. These barriers can include cultural stereotypes, misconceptions, and lack of awareness about neurodivergent conditions.

Cultural stereotypes and misconceptions about neurodivergent individuals can lead to stigma and discrimination, limiting their opportunities for education, employment, and social participation. For example, in some cultures, individuals with autism may be stigmatized as lacking intelligence or unable to contribute meaningfully to society, leading to exclusion and isolation.

Cultural norms and expectations can also influence attitudes and beliefs about seeking help and support for neurodivergent individuals. In some cultures, there may be a reluctance to acknowledge or discuss neurodivergent traits, which can limit access to diagnosis, interventions, and support services. This lack of awareness and understanding can perpetuate the marginalization of neurodivergent individuals in society.

The importance of culturally competent care and support

To promote acceptance and inclusion of neurodivergent individuals, it is crucial to develop culturally competent care and support systems. Culturally competent care acknowledges and respects the cultural perspectives and diversity within neurodivergent communities.

Culturally competent care involves understanding the cultural beliefs, values, and practices that influence the perception and treatment of neurodivergent traits. It requires adapting interventions and support strategies to align with the cultural context and needs of individuals and families.

For example, culturally competent care may involve collaborating with community leaders, educators, and healthcare providers to raise awareness about neurodivergent traits. It may also involve providing training and professional development to develop culturally sensitive assessment tools, diagnostic practices, and intervention approaches.

By incorporating cultural perspectives in the understanding and support of neurodivergent individuals, we can foster acceptance, reduce stigma, and promote the well-being of individuals from diverse cultural backgrounds.

Neurodiversity in Indigenous communities

Understanding neurodivergent traits within Indigenous communities requires recognizing the unique cultural contexts and perspectives. Indigenous communities often have their own ways of understanding and responding to neurodiversity that differ from mainstream Western approaches.

Indigenous cultures may view neurodivergent traits through a lens of interconnectedness and as part of a larger web of community and spiritual well-being. Neurodivergent individuals may be seen as possessing unique abilities and perspectives that contribute to the community's collective wisdom and resilience.

However, it is important to note that there is significant cultural and linguistic diversity among Indigenous communities worldwide. Each community has its own distinct practices, beliefs, and traditions, including the understanding and acceptance of neurodivergent traits. It is essential to engage in a collaborative and culturally sensitive approach when working with Indigenous communities to understand and support neurodivergent individuals.

Neurodiversity in global contexts

Neurodivergent traits are not limited to specific cultures or regions. They exist in diverse communities worldwide, and understanding neurodiversity within global contexts is vital in promoting acceptance and inclusion.

In some countries, government policies and initiatives actively support neurodivergent individuals' rights and well-being. These policies aim to provide equal opportunities for education, employment, and social participation. By recognizing neurodiversity as a valuable aspect of human diversity, these countries are challenging stigmatizing and exclusionary practices.

International collaboration and exchange of knowledge and best practices regarding neurodiversity are crucial for fostering greater acceptance and inclusion globally. By learning from different cultural perspectives, we can broaden our understanding of neurodivergent traits and enhance support systems for neurodivergent individuals across the world.

Summary

Cultural variations significantly influence the understanding and perception of neurodivergent traits. Different cultures may interpret behaviors associated with neurodiversity differently, impacting the diagnosis, treatment, and acceptance of neurodivergent individuals. Cultural barriers, such as stereotypes and misconceptions, can hinder acceptance and inclusion. To address these barriers, it is essential to develop culturally competent care and support strategies that respect and adapt to the cultural context and needs of neurodivergent individuals. Furthermore, recognizing neurodiversity within Indigenous communities and embracing global perspectives on neurodivergent traits are crucial for promoting acceptance and inclusion worldwide.

Cultural Barriers to Acceptance and Inclusion

Cultural barriers play a significant role in hindering the acceptance and inclusion of neurodivergent individuals in society. These barriers arise from deep-rooted beliefs, norms, and attitudes that are prevalent within specific cultures. Understanding and addressing these cultural barriers is essential to creating a more inclusive and supportive environment for neurodivergent individuals.

Beliefs and Attitudes towards Neurodivergent Traits

One of the primary cultural barriers to acceptance is the prevailing negative beliefs and attitudes towards neurodivergent traits. In many cultures, neurodivergent individuals are often stigmatized and seen as "different" or "abnormal." This stigma stems from a lack of understanding and awareness about neurodiversity, leading to misconceptions and stereotypes.

For example, in some cultures, individuals with autism may be perceived as "possessed by evil spirits" or "cursed." These beliefs can result in isolation, discrimination, and even harmful practices such as exorcism rituals. Similar negative attitudes and stereotypes may exist towards individuals with other neurodivergent conditions such as ADHD or dyslexia.

To address these cultural barriers, education and awareness campaigns are crucial. Promoting accurate information about neurodiversity and challenging misconceptions can help shift attitudes towards acceptance and inclusion. It is essential to engage communities and provide culturally sensitive resources that help dispel myths and foster understanding.

Cultural Norms and Expectations

Cultural norms and societal expectations can also act as barriers to acceptance and inclusion. In some cultures, there is an emphasis on conformity and fitting into established social norms. Neurodivergent individuals, who may exhibit behaviors or traits that deviate from these norms, may face difficulties in being accepted and included.

For instance, in certain cultures, there may be strong expectations for individuals to be highly social and outgoing. Neurodivergent individuals who struggle with social interactions and prefer solitude may be seen as outsiders or socially inept. This can lead to exclusion, marginalization, and a lack of inclusive opportunities.

Addressing these barriers requires a shift in cultural attitudes and an appreciation for diverse ways of being. Efforts should be made to challenge narrow definitions of "normal" and recognize the value of neurodivergent traits. By fostering acceptance of diverse cognitive profiles and redefining societal expectations, cultural norms can become more inclusive and supportive of neurodivergent individuals.

Communication and Language Barriers

Communication and language barriers can also limit acceptance and inclusion for neurodivergent individuals. Cultural differences in communication styles, norms, and languages can create challenges in understanding and connecting with neurodivergent individuals. This can lead to misunderstandings, misinterpretations, and even discrimination.

For example, neurodivergent individuals who communicate through nonverbal or unconventional means may face difficulties in cultures that prioritize verbal communication. Misjudgments and misinterpretations of their expressions and gestures can lead to exclusion and a lack of meaningful engagement.

To overcome these barriers, it is essential to promote inclusive communication practices. Culturally sensitive approaches should be adopted to accommodate different communication styles and to encourage active listening and understanding. Promoting the use of augmentative and alternative communication methods can also enhance inclusivity and enable effective communication with neurodivergent individuals.

Cultural Beliefs About Treatment and Interventions

Cultural beliefs about treatment and interventions can act as barriers to acceptance and inclusion. Some cultures may have traditional practices or beliefs that discourage seeking professional help or acknowledging neurodivergent conditions. This can lead to delayed or inadequate support for individuals who would benefit from appropriate interventions.

For instance, cultural beliefs that conditions like autism can be "cured" through alternative therapies or religious rituals may prevent individuals from accessing evidence-based interventions. These cultural barriers can impede early diagnosis, early intervention, and the provision of appropriate support that can enhance the well-being and quality of life for neurodivergent individuals.

It is crucial to engage with cultural communities to promote awareness of evidence-based practices and interventions. Collaborative efforts, involving healthcare professionals, educators, and community leaders, can help bridge the gap between cultural beliefs and effective support. Promoting culturally appropriate interventions and dispelling misconceptions about neurodivergent conditions can contribute to greater acceptance and inclusion.

Intersectionality and Cultural Barriers

Intersectionality refers to the interconnectedness of various social identities such as race, ethnicity, gender, and socioeconomic status. Cultural barriers to acceptance and inclusion are often influenced by intersectional factors, making it important to consider the multidimensionality of individuals' experiences.

For example, neurodivergent individuals who belong to marginalized communities may face additional barriers due to intersecting forms of discrimination and disadvantage. Discrimination based on both neurodivergent and cultural identities can result in further exclusion, limited opportunities, and a lack of access to necessary support services.

Addressing intersectionality requires taking an intersectional approach in efforts to promote acceptance and inclusion. This involves recognizing and addressing the unique challenges faced by individuals at the intersections of multiple identities. It involves creating inclusive spaces that celebrate diversity and account for the specific needs of neurodivergent individuals from diverse cultural backgrounds.

Conclusion

Cultural barriers to acceptance and inclusion pose significant challenges for neurodivergent individuals. Negative beliefs and attitudes, cultural norms and expectations, communication and language barriers, cultural beliefs about treatment and interventions, and intersectionality all contribute to these barriers. Overcoming these barriers requires comprehensive efforts that encompass education, awareness, and cultural sensitivity. By addressing cultural barriers, we can create a more inclusive and accepting society that values the diversity and potential of every individual, regardless of their neurodivergent traits.

The Importance of Culturally Competent Care and Support

Cultural competence is a critical component of providing effective care and support for neurodivergent individuals. It involves understanding and respecting the cultural beliefs, values, and practices that influence how individuals perceive and experience neurodiversity. Culturally competent care ensures that services are accessible, inclusive, and tailored to meet the diverse needs of individuals from different cultural backgrounds.

One key aspect of culturally competent care is recognizing the cultural variations in understanding and embracing neurodivergent traits. Different cultures have unique perspectives on neurodiversity, which can influence the level of acceptance and support available to neurodivergent individuals within their communities. For example, some cultures may view certain neurodivergent traits as a source of strength or heightened abilities, while others may stigmatize or pathologize these traits. Understanding these cultural variations is crucial in promoting acceptance, reducing stigma, and providing appropriate support to neurodivergent individuals.

Cultural barriers can also pose challenges to accessing care and support for neurodivergent individuals. Language barriers, limited cultural competence among service providers, and lack of awareness about available resources can hinder individuals from seeking the help they need. Culturally competent care addresses these barriers by providing language interpretation services, training providers to understand and respect cultural differences, and actively engaging with diverse communities to raise awareness of available support options.

Another important aspect of culturally competent care is the recognition of intersectionality. Neurodivergent individuals may belong to multiple marginalized groups, such as racial or ethnic minorities, LGBTQ+ communities, or individuals with disabilities. These intersecting identities can shape the experiences and needs of neurodivergent individuals, highlighting the importance of a holistic and culturally responsive approach to care and support.

To provide culturally competent care, professionals in the field must engage in ongoing education and self-reflection. This includes examining personal biases and assumptions, actively seeking training and resources to enhance cultural competence, and collaborating with community members and organizations to ensure services are aligned with cultural norms and values. It is also crucial to involve diverse perspectives in the decision-making processes related to policy development and service planning.

Practicing cultural competence goes beyond simply acknowledging cultural differences; it requires a shift in mindset and a commitment to inclusivity. This may involve adapting assessment tools, treatment approaches, and communication styles to accommodate diverse cultural backgrounds. It also involves building trust and establishing meaningful relationships with individuals and families by taking the time to understand their unique needs and experiences.

A real-world example of culturally competent care is the development of support programs specifically tailored to the needs of a particular cultural group. For instance, a community-based organization might establish a support group for neurodivergent individuals and their families within an Indigenous community.

The program would consider and integrate cultural practices, beliefs, and values into the support services, recognizing that Indigenous individuals may have unique perspectives on neurodiversity.

In conclusion, culturally competent care and support are essential for ensuring that the needs of neurodivergent individuals from diverse cultural backgrounds are met effectively. By recognizing cultural variations in understanding and embracing neurodiversity, addressing cultural barriers, and integrating diverse perspectives, professionals can provide inclusive and respectful care that promotes acceptance, reduces stigma, and empowers neurodivergent individuals to thrive within their own cultural contexts.

Neurodiversity in Indigenous Communities

Neurodiversity is not limited to specific cultures or regions but is a concept that transcends borders and embraces the unique strengths and challenges of individuals with neurodevelopmental conditions. In this section, we will explore the intersection of neurodiversity and indigenous communities, highlighting the cultural perspectives and practices related to neurodivergent traits.

Cultural Variations in Understanding Neurodivergent Traits

Indigenous communities around the world have diverse beliefs and understandings when it comes to neurodivergent traits. For many indigenous cultures, neurodivergent individuals are seen as having unique gifts and abilities rather than being labeled as "disabled" or "abnormal." The cultural lens through which neurodivergent traits are viewed is essential to understanding and appreciating the experiences of individuals within these communities.

In some indigenous cultures, for example, certain neurodivergent traits may be interpreted as signs of a spiritual connection or a heightened sensitivity to the environment. These traits are often revered and respected, with individuals being considered as having special roles within the community. The acceptance and inclusion of neurodivergent individuals are deeply ingrained in indigenous cultural practices, fostering a sense of belonging and value.

Cultural Barriers to Acceptance and Inclusion

It is important to acknowledge that not all indigenous communities uniformly embrace neurodivergent traits. Like any culture, indigenous communities can have variations in their understanding and acceptance of neurodiversity. Factors such as colonization, acculturation, and globalization have had an impact on traditional cultural practices and beliefs, including those related to neurodivergent individuals.

Historical trauma, forced assimilation, and the disruption of traditional ways of life have contributed to a loss of indigenous cultural practices that fostered acceptance and inclusion. As a result, some indigenous communities have adopted more medicalized or pathologizing approaches to neurodivergent traits, influenced by the dominant neurotypical perspective.

The Importance of Culturally Competent Care and Support

In order to promote neurodiversity in indigenous communities, it is crucial to develop culturally competent care and support systems. Culturally competent care recognizes and respects the unique cultural beliefs, practices, and values of indigenous communities while providing appropriate and inclusive services for neurodivergent individuals.

This includes fostering partnerships with elders, traditional healers, and community leaders to develop culturally appropriate interventions and support networks. It also involves incorporating traditional healing practices and incorporating indigenous knowledge systems into mainstream care approaches.

Indigenous communities have a wealth of knowledge and wisdom when it comes to holistic approaches to health and well-being. By valuing and integrating indigenous perspectives on neurodiversity, we can create a more inclusive and supportive environment for all individuals, regardless of their neurodivergent traits.

Neurodiversity in Indigenous Communities: An Example

To illustrate the cultural perspectives on neurodiversity in indigenous communities, let's consider the beliefs and practices of the Ojibwe people, an indigenous community in North America.

In Ojibwe culture, neurodivergent individuals are seen as having unique spiritual gifts and are often referred to as "spiritual warriors" or "keepers of sacred knowledge." These individuals are believed to possess a heightened sensitivity to the spiritual world and are valued for their abilities to connect with ancestors and spirits.

The Ojibwe have traditional ceremonies and practices that support the well-being of neurodivergent individuals. For example, the sweat lodge ceremony is used as a healing and purifying ritual, where individuals can connect with the spirit world and find balance and harmony within themselves. Drumming circles and storytelling are also used to provide a sense of belonging and community support.

By understanding and appreciating the Ojibwe perspective on neurodiversity, we can see the importance of cultural context and the role of spirituality in supporting the well-being of neurodivergent individuals. This example demonstrates the cultural diversity within indigenous communities and the need for culturally competent approaches to care and support.

Resources and Support

For individuals and professionals seeking to learn more about neurodiversity in indigenous communities, several resources and organizations provide valuable insights and support:

- Indigenous Disability and Neurodiversity Association: This organization aims to promote acceptance, inclusion, and empowerment of neurodivergent individuals within indigenous communities globally.
- "Diverse Indigenous Voices: Exploring the Intersection of Neurodiversity and Indigenous Knowledge" by Dr. Cheryl Kara-Seguine: This book explores the intersectionality between indigenous cultures and neurodivergent experiences, offering a deeper understanding of the unique cultural perspectives.
- Indigenous Healing Foundation: This organization focuses on healing and revitalizing indigenous communities through the recovery and reintegration of traditional health and wellness practices.
- Local indigenous community centers and organizations: Engaging with local community centers and organizations is an excellent way to learn directly from indigenous communities about their perspectives on neurodiversity and how to support neurodivergent individuals.

By accessing these resources and engaging in dialogue with indigenous communities, we can deepen our understanding of the cultural variations in understanding and embracing neurodivergent traits. This knowledge can help us develop more inclusive strategies and support systems for neurodivergent individuals within indigenous communities and beyond.

Conclusion

In this section, we explored the intersection of neurodiversity and indigenous communities, highlighting the cultural perspectives and practices related to neurodivergent traits. We learned that there is a wide variation in how neurodiversity is understood and accepted within indigenous cultures, and this understanding is crucial for providing appropriate care and support.

We emphasized the importance of culturally competent care and the need to incorporate indigenous knowledge systems into mainstream approaches. By valuing and integrating indigenous perspectives, we can create a more inclusive and supportive environment for all individuals, regardless of their neurodivergent traits.

Understanding neurodiversity in indigenous communities is an ongoing process that requires active engagement with indigenous peoples and organizations. By continuing to learn, listen, and collaborate, we can work towards a society that celebrates and embraces neurodiversity in all its forms.

Neurodiversity in Global Contexts

Neurodiversity is a concept that transcends cultural boundaries and is applicable in global contexts. Understanding neurodiversity from a global perspective involves recognizing the cultural variations in how neurodivergent traits are understood and the barriers that can hinder acceptance and inclusion. In this section, we will explore the importance of culturally competent care and support, the challenges faced by neurodivergent individuals in different cultural contexts, and the initiatives promoting neurodiversity in global communities.

Cultural Variations in Understanding Neurodivergent Traits

Neurodivergent traits manifest differently across cultures, and the understanding and acceptance of these traits can vary significantly. For example, in some cultures, traits associated with autism may be considered as indicators of "giftedness" or "spiritual enlightenment," while in others, they may be seen as signs of illness or abnormality. These cultural variations can influence the perception and treatment of neurodivergent individuals.

In some indigenous cultures, neurodivergent traits are valued and respected as unique abilities or talents that contribute to the community. For instance, in certain Native American tribes, people with neurodivergent traits are often regarded as visionary and are given important roles in spiritual ceremonies. Understanding and respecting these cultural perspectives is crucial for providing appropriate support and ensuring the inclusion of neurodivergent individuals.

Cultural Barriers to Acceptance and Inclusion

Despite the potential for positive cultural perspectives on neurodiversity, there are also cultural barriers that can hinder acceptance and inclusion. Stigma, lack of awareness, and cultural norms surrounding disability can create additional challenges for neurodivergent individuals and their families. In some cultures, there may be a tendency to hide or deny neurodivergent traits due to shame or fear of social discrimination.

Cultural attitudes towards mental health and disability can also have a significant impact. In cultures where mental health issues and disabilities are highly stigmatized, seeking help or accessing support services may be seen as a sign of weakness or failure. This can further isolate neurodivergent individuals and limit the availability of resources and support networks.

The Importance of Culturally Competent Care and Support

Culturally competent care and support play a crucial role in promoting neurodiversity in global contexts. This involves understanding the cultural beliefs, values, and practices that shape the perception and treatment of neurodivergent individuals within specific communities. Culturally competent practitioners are able to provide tailored interventions and support that respect and accommodate cultural differences.

Collaboration with community leaders and stakeholders is vital for implementing culturally competent care. Engaging with local communities and involving them in the development of support programs can help address cultural barriers and ensure that interventions are meaningful and effective. It is essential to establish trust and build relationships with families and community members to facilitate the acceptance and inclusion of neurodivergent individuals.

Neurodiversity in Indigenous Communities

Indigenous communities around the world have unique perspectives on neurodiversity rooted in their cultural heritage. These communities often view neurodivergent traits as sacred or shamanic gifts that contribute to the overall well-being of the community. However, colonization, cultural assimilation, and modernization have disrupted traditional practices and led to the marginalization of indigenous people, including those with neurodivergent traits.

Efforts are being made to reclaim and revitalize indigenous cultures, including recognizing and embracing neurodivergent individuals within these communities. Indigenous-led initiatives focus on preserving cultural practices, promoting holistic well-being, and creating inclusive spaces for neurodivergent individuals to thrive. These initiatives highlight the importance of culturally competent approaches that integrate traditional knowledge and values.

Neurodiversity in Global Initiatives

The recognition of neurodiversity as a global phenomenon has led to the emergence of global initiatives and networks advocating for the rights and well-being of neurodivergent individuals. These initiatives aim to foster cross-cultural dialogue, share best practices, and promote the principles of neurodiversity worldwide.

International organizations such as the Autistic Self Advocacy Network and the AsIAm Foundation actively work to amplify the voices of neurodivergent individuals and promote inclusion and acceptance. These organizations engage in advocacy, raise awareness, and provide resources for individuals and communities around the world to support neurodiversity.

Global conferences and events focused on neurodiversity bring together individuals, researchers, and practitioners from diverse cultural backgrounds. These platforms facilitate the exchange of knowledge, experiences, and innovative approaches to promoting neurodiversity in different cultural contexts. They also provide opportunities for collaboration and solidarity across borders.

Promoting Neurodiversity in a Globalized World

In a globalized world, where cultures interact and influence one another, promoting neurodiversity requires a holistic approach that transcends cultural boundaries. It calls for embracing cultural diversity, challenging stereotypes, and fostering mutual respect and understanding.

Education systems should incorporate culturally relevant curricula that value neurodiversity and promote inclusive practices. This includes training teachers and educational staff to understand and support neurodivergent students from various cultural backgrounds. It also involves promoting intercultural awareness and empathy among students, fostering a more inclusive and accepting learning environment.

Furthermore, policies and legislation should reflect the principles of neurodiversity and address the specific needs of neurodivergent individuals within different cultural contexts. This involves collaborating with local communities and advocacy groups to ensure that the voices and concerns of neurodivergent individuals are heard and addressed.

In conclusion, neurodiversity in global contexts encompasses the cultural variations in understanding and accepting neurodivergent traits. While cultural perspectives on neurodiversity can be both supportive and challenging, promoting neurodiversity requires culturally competent care and support. Initiatives and networks at the global level play a crucial role in raising awareness, fostering cross-cultural dialogue, and promoting the rights and well-being of neurodivergent individuals worldwide. By embracing cultural diversity and challenging stereotypes, we can work towards a more inclusive and accepting world that celebrates neurodiversity.

Education and Neurodiversity

Inclusive Education

The Principles and Benefits of Inclusive Education

Inclusive education is a philosophy and approach that promotes the full participation and equal opportunities for all learners, regardless of their abilities, disabilities, or neurodivergent traits. It is based on the principle that every individual has the right to access quality education in a supportive and inclusive environment. Inclusive education goes beyond mere integration, where students with disabilities are placed in regular classrooms without necessary adjustments or supports. Instead, it aims to create a truly inclusive learning community that values and embraces diversity.

The benefits of inclusive education are manifold. Firstly, it fosters a sense of belonging and acceptance among all learners. When students with diverse abilities and backgrounds come together in the same classroom, it promotes understanding, empathy, and respect for differences. This inclusive environment nurtures a positive social and emotional climate, reducing the risk of bullying, exclusion, and stigmatization. It helps cultivate a culture of diversity and inclusion, not only within the school but also in society at large.

Inclusive education also maximizes learning opportunities for all students. By recognizing and addressing the individual needs and strengths of each student, inclusive classrooms provide tailored support and accommodations. This ensures that every learner can access the curriculum and actively participate in learning activities. Teachers employ a variety of instructional strategies, including differentiated instruction, multi-sensory materials, and assistive technologies, to meet diverse learning styles and needs. Inclusive education is not only beneficial for students with disabilities or neurodivergent traits but also for neurotypical students, who benefit from exposure to different perspectives and learning strategies.

Moreover, inclusive education promotes academic achievement and long-term educational outcomes. Research consistently shows that when students with disabilities are educated in inclusive settings, they demonstrate better academic performance, higher graduation rates, and improved post-secondary transitions. Inclusive classrooms provide opportunities for collaboration and peer learning, which enhances critical thinking, problem-solving skills, and creativity.

Inclusive education also has positive effects on the social skills and self-esteem of students. Through interactions with their peers and teachers, students learn to appreciate differences, develop empathy, and become more tolerant. They build friendships and supportive networks that extend beyond the classroom, contributing to their overall well-being and sense of belonging. Inclusive education helps students develop a positive self-concept and self-advocacy skills, empowering them to advocate for their needs and rights in all aspects of life.

To achieve inclusive education, several key principles need to be embraced and implemented. These include:

1. Universal Design for Learning (UDL): UDL provides a framework for creating inclusive learning environments by offering multiple means of representation, engagement, and expression. It ensures that instructional materials, activities, and assessments are accessible to all learners, catering to their individual strengths and needs.

2. Individualized Education Plans (IEPs) and Supports: Individualized education plans are essential for students with disabilities or neurodivergent traits. They outline specific goals, accommodations, and support services that students require to achieve their educational outcomes. IEPs are developed collaboratively with the input of students, parents, teachers, and other professionals.

3. Differentiated Instruction: Differentiated instruction is a teaching approach that recognizes and accommodates the diverse learning needs and preferences of students. It involves adapting instruction, content, and assessment to meet the individual needs, interests, and readiness levels of learners.

4. Collaborative and Multidisciplinary Approaches: Inclusive education involves collaboration and partnership between teachers, support staff, parents, and community professionals. This collaborative approach ensures that all stakeholders work together to support the holistic development and learning of students.

5. Inclusive Classroom Environment: Creating an inclusive classroom environment involves establishing clear expectations, fostering positive relationships, and promoting a culture of respect and acceptance. It includes

creating physical spaces that are accessible and flexible, allowing for different learning styles and needs.

Inclusive education is not without its challenges. It requires adequate funding, resources, and ongoing professional development for teachers. It necessitates systemic changes in educational policies, practices, and attitudes. However, the benefits of inclusive education far outweigh the challenges. It is a powerful tool for promoting social justice, equal opportunities, and the overall well-being of all learners.

In conclusion, inclusive education is a transformative approach that celebrates the diversity of learners and promotes their full participation in educational and social contexts. It embraces the principles of equity, accessibility, and individualized support. By fostering inclusive learning communities, we can create a future that values and celebrates the unique strengths and abilities of every individual, regardless of their neurodivergence.

Universal Design for Learning (UDL)

In the field of education, Universal Design for Learning (UDL) is a framework that aims to ensure that all students have equal opportunities to learn, regardless of their diverse learning needs, abilities, and backgrounds. UDL is based on the principles of universal design, which originated in architecture and product design and emphasizes creating environments that are accessible to and beneficial for all individuals. The concept of UDL can be applied to instructional practices, curriculum development, and assessment strategies to create more inclusive learning experiences for students.

Background

UDL was first introduced by the Center for Applied Special Technology (CAST) in the 1990s. The framework was developed as a response to the recognition that traditional teaching methods often failed to meet the needs of students with disabilities and diverse learning styles. UDL aims to provide a more flexible and personalized approach to education, enabling every student to access and engage with the curriculum effectively.

Principles of UDL

UDL is guided by three core principles, collectively known as the UDL guidelines:

1. **Multiple Means of Representation:** This principle focuses on providing various ways of presenting information to students. By offering content in multiple formats, such as text, audio, visuals, or multimedia, educators can cater to different learning preferences and ensure that all students can access and comprehend the material effectively.

   For example, a teacher might provide a video lecture, written notes, and a discussion forum for students to engage with the same topic.

2. **Multiple Means of Action and Expression:** This principle emphasizes allowing students to demonstrate their knowledge and skills in different ways. It recognizes that students have diverse strengths and preferences when it comes to expressing what they have learned. Providing options for different modes of expression, such as written assignments, presentations, or multimedia projects, allows students to showcase their understanding in a way that aligns with their abilities and interests.

   For instance, a student with dyslexia may choose to create a visual poster instead of writing an essay to demonstrate their understanding of a concept.

3. **Multiple Means of Engagement:** This principle aims to enhance student motivation and engagement by offering various ways to connect with the curriculum. Different students are motivated by different aspects of learning, and UDL encourages educators to provide options that cater to diverse interests and preferences. This can be achieved by offering choices in learning materials, providing opportunities for collaboration, and incorporating real-world examples and authentic tasks into the curriculum.

   For example, a science teacher might allow students to choose between conducting a hands-on experiment, watching a documentary, or researching a scientific case study to explore a specific topic.

Implementing UDL in Practice

Implementing UDL in practice involves adopting a flexible and proactive approach to instructional design and delivery. Here are some strategies and considerations for incorporating UDL principles into teaching:

- **Assessment:** Provide multiple assessment options that allow students to demonstrate their understanding in different ways. This could include traditional tests, project-based assessments, oral presentations, or multimedia portfolios. Consider using rubrics that focus on essential understanding and skills rather than specific formats or techniques.

- **Flexible Materials:** Offer learning materials in various formats, such as written text, audio recordings, videos, or interactive online modules. Use accessible design principles, such as providing captions and transcripts for multimedia content and ensuring readability and clarity in written materials.

- **Instructional Design:** Incorporate a variety of instructional strategies and methods to cater to different learning preferences and needs. This could involve using explicit instruction, hands-on activities, problem-solving tasks, collaborative group work, or online interactive tools. Provide clear learning goals and objectives and offer multiple entry points to help students build on their existing knowledge and skills.

- **Technology Integration:** Utilize technology tools and resources to support UDL implementation. Digital platforms and assistive technologies can offer customizable options for students, including text-to-speech, speech-to-text, adjustable font sizes, and interactive multimedia components. Consider providing students with choices of technology tools based on their individual needs and preferences.

- **Collaboration and Communication:** Encourage collaboration and peer interaction among students to foster a supportive learning environment. Provide opportunities for group work, discussions, and feedback exchanges. Foster a culture of respect and empathy, where students can learn from each other's diverse perspectives and experiences.

- **Ongoing Reflection and Evaluation:** Regularly reflect on the effectiveness of UDL strategies and seek feedback from students to make improvements. Evaluate student progress based on individual growth and achievement rather than rigid benchmarks. Engage in professional development opportunities to expand knowledge and skills related to UDL implementation.

Benefits and Challenges of UDL

The adoption of UDL principles offers numerous benefits for both students and educators. By providing multiple means of representation, action and expression, and engagement, UDL promotes equity, inclusivity, and meaningful learning experiences. UDL supports the removal of barriers to learning, fosters student independence and self-advocacy skills, and promotes the development of a growth mindset among students.

However, implementing UDL can also present challenges. Educators may face constraints in terms of time, resources, and support. Adapting instructional materials and utilizing assistive technologies, although beneficial, requires additional planning and preparation. Collaborating with colleagues, involving families, and connecting with community resources can be essential in overcoming these challenges and maximizing the potential of UDL implementation.

Conclusion

Universal Design for Learning is a powerful framework that aims to transform the learning experience for all students. By embracing the principles of UDL, educators can create inclusive and engaging learning environments that honor the diversity of learners. Through the provision of multiple means of representation, action and expression, and engagement, UDL enables students to reach their full potential and become active, motivated, and independent learners.

As you begin to explore the application of UDL in your own teaching practice, remember to consider the unique needs and strengths of your students, seek feedback and input from them, and continuously reflect on and refine your instructional strategies. By doing so, you will contribute to the creation of a truly inclusive and empowering educational experience for all students.

Individualized Education Plans (IEPs) and Supports

Inclusive education aims to provide equal opportunities for all learners, including those with neurodivergent conditions. One important tool in achieving this goal is the Individualized Education Plan (IEP). An IEP is a legally binding document that outlines the specific educational goals and support services for a student with special needs. It is a collaborative process that involves the student, their parents or caregivers, teachers, and other professionals.

Understanding the Purpose of IEPs

The primary purpose of an IEP is to tailor the educational program to meet the unique needs of the individual student. It recognizes that one size does not fit all and acknowledges that students with neurodivergent conditions require personalized support to thrive academically, socially, and emotionally. The IEP ensures that the educational goals are aligned with the student's strengths, challenges, and interests.

Components of an IEP

An IEP typically includes several components that together form a comprehensive plan for supporting the student. These components may vary depending on the specific needs of the student but commonly include the following:

- **Present Level of Performance:** This section describes the student's current academic and functional abilities, highlighting their strengths and areas of need. It serves as a baseline for setting goals and measuring progress.

- **Goals and Objectives:** IEP goals are specific, measurable, achievable, relevant, and time-bound (SMART). They reflect the areas in which the student requires support and outline the expected progress over a specific time frame. Objectives are specific steps towards achieving the goals.

- **Special Education and Related Services:** This section identifies the specialized services the student will receive, such as speech therapy, occupational therapy, or behavioral interventions. It specifies the frequency, duration, and location of these services.

- **Accommodations and Modifications:** Accommodations are changes to the learning environment that allow the student to access the curriculum on an equal basis. Modifications, on the other hand, involve altering the curriculum itself. Both accommodations and modifications should be tailored to the student's needs and promote their participation and success.

- **Transition Planning:** For older students, the IEP includes transition planning, which focuses on preparing the student for life beyond school. It may involve career exploration, vocational training, college planning, or independent living skills development.

- **Parent and Student Participation:** The IEP process emphasizes the involvement of parents or caregivers and the student whenever appropriate. Their input and perspectives are vital in developing a meaningful and effective plan. They are also partners in monitoring progress and making adjustments along the way.

Developing and Implementing an IEP

The process of developing and implementing an IEP involves collaboration and ongoing communication between all stakeholders. It typically follows these steps:

1. **Referral and Evaluation:** A student is referred for assessment if there are concerns about their educational progress or suspected disabilities. Various assessments, such as cognitive testing, academic evaluations, or behavioral observations, may be conducted to gather information about the student's strengths and challenges.

2. **IEP Meeting:** Once the evaluation is complete and the student is found eligible for special education services, an IEP meeting is scheduled. This meeting brings together parents, teachers, specialists, and the student (when appropriate) to develop the IEP. The team reviews assessment results, sets goals, and determines the necessary supports and services.

3. **Written IEP:** Based on the discussions and decisions made during the IEP meeting, a written IEP is developed. It includes all the components mentioned earlier and is reviewed and revised annually or more frequently if needed.

4. **IEP Implementation:** Once the IEP is finalized, the student's teachers and support staff receive a copy and are responsible for implementing the plan. Regular monitoring of progress and support adjustments are crucial during this phase.

5. **Progress Monitoring and Reporting:** Progress towards the goals outlined in the IEP is regularly monitored, and periodic reports are shared with parents and caregivers. This monitoring helps identify challenges, determine the effectiveness of strategies, and make any necessary modifications.

Challenges and Considerations

While IEPs are designed to support students with neurodivergent conditions, there are still challenges and considerations to be aware of:

- **Ensuring Individualization:** It is essential that IEPs truly reflect the individual needs of students. This requires thorough assessments, meaningful goals, and appropriate accommodations and services. Avoiding a one-size-fits-all approach is crucial for the success of inclusive education.

- **Collaboration and Communication:** Effective collaboration and ongoing communication among all stakeholders are key to the successful implementation of an IEP. Clear lines of communication, regular progress meetings, and an inclusive partnership approach help ensure that everyone is working towards the same goals.

- **Transition Planning:** Transition planning in an IEP should be comprehensive and begin early. It should address not only academic and vocational goals but also the student's social and emotional readiness for life after school. Engaging community resources and establishing connections with potential employers or post-secondary institutions can greatly support the transition process.
- **Advocacy and Empowerment:** Students with neurodivergent conditions should be actively involved in the IEP process as it fosters self-advocacy and empowers them to take ownership of their education. Encouraging students to voice their opinions, concerns, and goals helps create a sense of agency and promotes self-determination.

Conclusion

Individualized Education Plans (IEPs) play a vital role in supporting the educational journey of neurodivergent students. Through collaborative efforts and personalized strategies, IEPs enable students to access the curriculum, develop important skills, and thrive in inclusive educational environments. By recognizing the uniqueness of each student and embracing the principles of neurodiversity, IEPs contribute to creating equitable and inclusive educational opportunities for all learners. Remember, the true success of an IEP lies in its ability to empower neurodivergent students to reach their full potential and celebrate their individual strengths.

Differentiated Instruction and Teaching Strategies

In the inclusive classroom, one size does not fit all. Students come from diverse backgrounds, possess different learning styles, and have unique abilities and interests. To ensure that all students can thrive and reach their full potential, educators need to employ differentiated instruction and teaching strategies. This approach recognizes and values the individuality of each student and tailors instruction to meet their specific needs. In this section, we will explore the principles of differentiated instruction, effective strategies, and the benefits it brings to diverse learners.

Principles of Differentiated Instruction

Differentiated instruction is rooted in the belief that every student can succeed when provided with appropriate support and opportunities. It is guided by the following principles:

1. **Flexibility:** Differentiated instruction offers flexibility in content, process, and product. It allows students to access the curriculum in ways that are meaningful to them and demonstrate their understanding through multiple modes of expression.

2. **Individualization:** This approach acknowledges the diversity of learners and individualizes instruction based on their unique strengths, interests, and needs. It recognizes that students may require different support systems to succeed.

3. **Assessment for Learning:** Ongoing and varied assessments are used to gather data about students' progress and interests. This information informs instruction and allows teachers to adapt their approaches to meet individual needs effectively.

4. **Collaboration:** Differentiated instruction promotes collaboration among students, teachers, and families. It encourages students to learn from each other, share ideas, and work in groups to achieve common goals.

Effective Strategies for Differentiated Instruction

Implementing differentiated instruction requires thoughtful planning and a variety of instructional strategies. Here are some effective strategies that can be used to differentiate instruction:

1. **Tiered Assignments:** Assignments are created at different levels of complexity or depth. Students can choose assignments that best match their abilities and learning preferences while still addressing the same essential concepts.

2. **Learning Centers/Stations:** Different learning activities are set up at various stations or centers in the classroom. Students can rotate through these stations, engaging in tasks that cater to their specific needs or interests.

3. **Flexible Grouping:** Students are grouped based on their readiness, interests, or learning profiles. Group composition is flexible and may change depending on the task or learning goal.

4. **Varied Materials and Resources:** Provide students with a range of materials such as texts, multimedia, manipulatives, and technology tools to accommodate diverse learning styles and preferences.

5. **Scaffolded Instruction:** Break down complex tasks into smaller, manageable steps. Provide appropriate support to help students build their skills and knowledge gradually.

6. **Alternative Assessment Methods:** Offer students different ways to demonstrate their understanding, such as presentations, projects, portfolios, or performances. This allows students to showcase their strengths and talents in alternative ways.

Benefits of Differentiated Instruction

Differentiated instruction offers numerous benefits for both students and teachers:

1. **Increased Student Engagement:** When instruction is tailored to students' interests and needs, they are more engaged and motivated to learn. This leads to better retention of information and deeper understanding of concepts.

2. **Improved Academic Achievement:** By meeting students where they are and providing appropriate support, differentiated instruction promotes academic growth and success. It allows students to progress at their own pace and reach their full potential.

3. **Positive Learning Environment:** Differentiated instruction creates an inclusive and supportive classroom environment where all students feel valued and respected. It fosters a sense of belonging and promotes positive relationships among students.

4. **Enhanced Teacher Satisfaction:** Through differentiated instruction, teachers can witness the progress and growth of each student. It allows them to tap into their creativity and adapt their teaching methods to meet individual needs, leading to a greater sense of professional satisfaction.

It is important to note that implementing differentiated instruction requires ongoing reflection, collaboration, and adjustment. Teachers need to continuously assess student needs and gather feedback to refine their instructional practices. By embracing differentiated instruction, educators can create a classroom environment that celebrates diversity and maximizes the potential of all learners.

Example: Let's consider a math lesson on fractions. In a differentiated classroom, the teacher might provide different activities for students with varying levels of understanding. Some students may work on basic fraction concepts, while others tackle more advanced problems. The teacher might use manipulatives, such as fraction bars or circles, to support visual learners. Additionally, students could choose to work individually or in small groups, depending on their preference. By differentiating the instruction, each student is given the opportunity to grow and succeed in their understanding of fractions.

Resources: - Tomlinson, C. (2017). *How to Differentiate Instruction in Academically Diverse Classrooms.* ASCD. - Gregory, G. H., & Chapman, C. (2013). *Differentiated Instructional Strategies: One Size Doesn't Fit All.* Corwin.

Key Takeaways: - Differentiated instruction personalizes the learning experience based on students' strengths, interests, and needs. - Flexibility, individualization, assessment for learning, and collaboration are key principles of differentiated instruction. - Strategies such as tiered assignments, learning centers, flexible grouping, and varied materials support differentiated instruction. -

Differentiated instruction leads to increased student engagement, improved academic achievement, a positive learning environment, and enhanced teacher satisfaction.

Collaborative and Multidisciplinary Approaches in Education

In the field of education, collaborative and multidisciplinary approaches have gained recognition for their effectiveness in supporting neurodivergent learners. These approaches recognize that addressing the diverse needs of neurodivergent students requires the collaboration and involvement of various professionals, including educators, therapists, counselors, and parents. This section will explore the principles and benefits of collaborative and multidisciplinary approaches, discuss key strategies for successful implementation, and provide examples of how these approaches can enhance educational outcomes for neurodivergent students.

Principles of Collaborative and Multidisciplinary Approaches

Collaborative and multidisciplinary approaches are rooted in the belief that no single professional can fully meet the complex needs of neurodivergent learners. By working together, professionals from different disciplines can contribute their unique expertise and perspectives, resulting in more comprehensive and holistic support for students. The following principles guide the implementation of collaborative and multidisciplinary approaches:

1. **Shared goals and objectives:** All professionals involved in the collaborative team must have a common understanding of the educational goals and objectives for the neurodivergent student. By establishing shared goals, the team can work cohesively towards the student's success.

2. **Effective communication and information sharing:** Regular and clear communication among team members is essential for successful collaboration. This includes sharing relevant information about the student's strengths, challenges, progress, and any interventions or accommodations that have been implemented.

3. **Respect for diverse expertise:** Each team member brings their own unique knowledge and skills to the collaborative process. It is important to recognize and respect the expertise of all professionals involved, fostering a culture of mutual learning and growth.

4. **Interdisciplinary problem-solving:** Collaborative teams engage in problem-solving discussions to address the unique challenges faced by neurodivergent students. Through interdisciplinary discussions, the team can develop creative and effective solutions that draw on the expertise of each member.

5. **Continuity of support:** The collaborative team strives to ensure that support for the neurodivergent student is consistent across different settings and transitions. This may involve the coordination of services between school, home, and community environments to provide a unified and seamless support system.

Strategies for Successful Collaboration

Implementing collaborative and multidisciplinary approaches requires careful planning and coordination among professionals. The following strategies can facilitate successful collaboration:

1. **Regular team meetings:** Establishing regular team meetings allows professionals to come together to discuss the student's progress, evaluate the effectiveness of interventions, and make adjustments as needed. These meetings also provide an opportunity for information sharing and collaborative problem-solving.

2. **Clear roles and responsibilities:** Clearly defining the roles and responsibilities of each team member helps ensure that everyone understands their contributions to the collaborative process. This clarity enables effective coordination and avoids duplication of efforts.

3. **Shared documentation and data-tracking:** Centralized documentation and data-tracking systems enable professionals to access and update relevant information about the student's progress in real-time. This promotes transparency and ensures that all team members are working with the most up-to-date information.

4. **Professional development and training:** Providing professional development opportunities for team members helps enhance their knowledge and skills in supporting neurodivergent students. This can include training on evidence-based practices, understanding the impact of neurodivergence on learning, and effective collaboration strategies.

5. **Parent and caregiver involvement:** Collaboration extends beyond professionals to include parents and caregivers as integral members of the team. Involving parents and caregivers in decision-making processes and providing them with the necessary support and resources strengthens the overall support system for the neurodivergent student.

Benefits of Collaborative and Multidisciplinary Approaches

Collaborative and multidisciplinary approaches in education offer a range of benefits for neurodivergent students:

1. **Holistic support and individualized interventions:** By bringing together professionals from different disciplines, collaborative teams can develop individualized interventions that address the specific strengths and challenges of each neurodivergent student. This holistic approach takes into account the cognitive, social, emotional, and behavioral aspects of the student's development.

2. **Increased accountability and continuity:** With professionals working together and sharing information, there is a higher level of accountability for the student's progress. The continuity of support across different settings and transitions ensures a consistent and coherent approach to meeting the student's needs.

3. **Optimal use of resources:** Collaborative and multidisciplinary approaches maximize the efficient use of available resources. By pooling together their expertise, professionals can identify and implement interventions that are evidence-based, cost-effective, and tailored to the student's unique needs.

4. **Enhanced problem-solving and innovation:** The diverse perspectives and expertise within collaborative teams foster creative problem-solving and innovation. Professionals can draw on their collective knowledge to develop new and effective strategies that promote the neurodivergent student's success.

5. **Empowerment and advocacy:** Collaborative teams can empower neurodivergent students by involving them in the decision-making process and helping them develop self-advocacy skills. Through collaborative approaches, students become active participants in their education and are better equipped to navigate challenges and advocate for their needs.

Example: Collaborative Approach for a Student with Autism Spectrum Disorder (ASD)

To illustrate the practical application of collaborative and multidisciplinary approaches, let's consider an example of a student with Autism Spectrum Disorder (ASD). In this case, the collaborative team may include the following professionals: special education teacher, speech therapist, occupational therapist, school psychologist, and the student's parents.

The collaborative team meets regularly to discuss the student's progress, share observations, and review assessment data. They work together to develop an individualized education plan (IEP) that includes specific goals and objectives tailored to the student's needs. The special education teacher implements differentiated instruction strategies in the classroom, while the speech therapist and occupational therapist provide targeted interventions to address the student's speech and sensory-motor challenges. The school psychologist conducts assessments and provides counseling support to address the student's social and emotional well-being.

The collaborative team also involves the student's parents, ensuring open communication and active involvement in decision-making. The team shares resources and strategies with the parents, empowering them to support their child's learning and development at home.

Through this collaborative and multidisciplinary approach, the student with ASD receives comprehensive support that addresses their unique strengths and challenges. The collaboration among professionals ensures a holistic and coordinated approach to meeting the student's educational needs, leading to improved outcomes and a positive educational experience.

Conclusion

Collaborative and multidisciplinary approaches in education are essential for effectively supporting neurodivergent learners. By bringing together professionals from different disciplines, these approaches foster a comprehensive, individualized, and holistic approach to education. The principles of shared goals, effective communication, respect for diverse expertise, interdisciplinary problem-solving, and continuity of support guide the implementation of these approaches. The benefits include holistic support, increased accountability, optimal resource utilization, enhanced problem-solving and innovation, and empowerment of neurodivergent students. Through collaborative and multidisciplinary approaches, we can create inclusive educational environments that celebrate neurodiversity and ensure the success of all learners.

Neurodivergent Success Stories in Education

Advocacy and Empowerment in Education

Advocacy and empowerment are critical components of creating an inclusive and supportive educational environment for neurodivergent students. By advocating for their rights and promoting their empowerment, we can ensure that these students have equal access to educational opportunities and can reach their full potential. In this section, we will explore the importance of advocacy and empowerment in education, discuss strategies for promoting these principles, and provide examples of successful initiatives.

Understanding Advocacy in Education

Advocacy in education refers to the process of supporting and speaking up for the rights and needs of neurodivergent students. It involves ensuring that their voices are heard, their needs are met, and their rights are protected. Advocacy can take many forms, including individual advocacy, where parents or caregivers advocate on behalf of their child, and systemic advocacy, where organizations and professionals work to improve educational policies and practices for all neurodivergent students.

Advocacy plays a crucial role in addressing the barriers and challenges faced by neurodivergent students in education. By raising awareness about neurodiversity and promoting inclusive practices, advocates can help to change the narrative surrounding these students and challenge the stereotypes and misconceptions that exist.

Promoting Empowerment in Education

Empowerment is closely linked to advocacy and involves helping neurodivergent students develop self-confidence, self-advocacy skills, and a sense of agency. Empowering these students means enabling them to take an active role in their education and decision-making processes. It involves creating an environment that values their strengths, interests, and unique perspectives.

There are several strategies for promoting empowerment in education. One important approach is fostering self-advocacy skills by providing opportunities for neurodivergent students to express their needs and preferences. This can be done through individualized education plans (IEPs) and self-determination programs, which encourage students to set goals, make choices, and participate in planning their education.

Additionally, educators can promote empowerment by creating a supportive and inclusive classroom environment. This includes fostering a positive and accepting culture, encouraging collaboration and peer support, and providing opportunities for students to showcase their strengths and talents. By providing accommodations and tailored support, educators can ensure that neurodivergent students feel valued and included.

Successful Initiatives in Advocacy and Empowerment

Numerous successful initiatives have been implemented to promote advocacy and empowerment in education for neurodivergent students. One example is the Autism Self-Advocacy Network (ASAN), a self-advocacy organization run by and for individuals on the autism spectrum. ASAN provides resources, training, and support to foster self-advocacy skills and empower autistic individuals to become leaders in their communities and shape policies that affect their lives.

Another successful initiative is the inclusion of neurodivergent students in the development of educational policies and practices. By actively involving neurodivergent students in decision-making processes, schools and organizations can ensure that their perspectives are considered and their needs are met. This can be achieved through student-led committees, focus groups, and consultation sessions.

Furthermore, partnerships between schools, families, and community organizations can be instrumental in promoting advocacy and empowerment in education. Collaborative efforts can provide a platform for sharing knowledge, resources, and experiences, while also strengthening the collective voice of neurodivergent individuals and their allies.

Challenges and Solutions

Although advocacy and empowerment are essential for the success of neurodivergent students in education, there are challenges that need to be addressed. One challenge is the limited awareness and understanding of neurodiversity among educators and the broader community. To overcome this challenge, professional development programs can be implemented to increase awareness and provide educators with strategies for creating inclusive learning environments.

Another challenge is the need for systemic changes in educational policies and practices to promote inclusivity. This requires advocacy at the policy level to ensure that laws and regulations support the rights and needs of neurodivergent students. It also involves implementing inclusive practices, such as universal design for learning (UDL), that benefit all students, regardless of their neurodivergent status.

Additionally, the lack of resources and funding can pose a challenge in promoting advocacy and empowerment in education. To address this, schools and organizations can seek partnerships with community stakeholders, apply for grants, and allocate resources specifically for promoting neurodiversity inclusion.

Summary

Advocacy and empowerment are fundamental for creating an inclusive and supportive educational environment for neurodivergent students. Through advocacy, we can raise awareness, challenge stereotypes, and ensure that the rights and needs of neurodivergent students are protected. Empowerment, on the other hand, involves fostering self-confidence, self-advocacy skills, and a sense of agency in neurodivergent students. Together, advocacy and empowerment promote equal access, inclusivity, and success for all students in education. By implementing strategies, addressing challenges, and learning from successful initiatives, we can create educational environments that celebrate neurodiversity and provide opportunities for all students to thrive.

Specialized Programs and Supports for Neurodivergent Learners

Specialized programs and supports play a crucial role in promoting the success and well-being of neurodivergent learners. These programs provide tailored interventions, accommodations, and resources that address the unique needs of individuals with neurodevelopmental conditions. In this section, we will explore some of the key specialized programs and supports that have been proven effective in supporting neurodivergent learners.

Individualized Education Plan (IEP)

One of the fundamental components of supporting neurodivergent learners is the development and implementation of an Individualized Education Plan (IEP). An IEP is a personalized document that outlines the goals, services, accommodations, and supports to meet the specific educational needs of a neurodivergent student. It is a collaborative effort involving educators, families, and other professionals.

The IEP process begins with the identification of the student's strengths, needs, and areas of challenge. This information is used to set meaningful and attainable goals that align with the student's unique abilities. The plan also specifies the supports and accommodations that will be provided, such as extra time on assignments, preferential seating, visual prompts, or access to assistive technology.

Regular assessment and monitoring are essential in ensuring that the IEP is effective and that progress is being made towards the identified goals. Through ongoing collaboration and communication among all stakeholders, the IEP can be adjusted and refined to meet the changing needs of the neurodivergent learner.

Specialized Instructional Programs

Specialized instructional programs offer targeted interventions and strategies designed to address the specific learning profiles of neurodivergent learners. These programs focus on building foundational skills, such as social communication, emotional regulation, executive functioning, and academic competencies.

For example, social skills training programs provide neurodivergent learners with opportunities to develop and practice effective social communication strategies. These programs often use structured teaching methods, role-playing exercises, and real-life simulations to enhance social understanding, perspective-taking, and problem-solving abilities.

Another example is the executive functioning skills programs, which assist neurodivergent learners in developing organization, time management, planning, and self-monitoring skills. These programs may include the use of visual supports, schedules, checklists, and strategies to improve task initiation, attention, and goal-directed behaviors.

Specialized instructional programs are typically implemented within the school setting and may involve small group instruction, one-on-one support, or a combination of both. These programs are tailored to the individual needs of the student to optimize learning and promote success across various domains.

Transition Programs

Transition programs aim to support neurodivergent learners as they transition from one educational level to another or from school to post-secondary education, employment, or independent living. These programs recognize the unique challenges faced by neurodivergent individuals during these transitional periods and provide the necessary resources, skills, and support to facilitate a successful transition.

For instance, transition programs may offer career exploration and vocational training opportunities to prepare neurodivergent learners for the workforce. These programs focus on identifying the individual's strengths, interests, and abilities and connecting them with appropriate career pathways. Additionally, they may provide job coaching, internships, and supported employment services to ease the transition into the workplace.

Transition programs also address the social and emotional aspects of transitioning by offering individual and group counseling, peer support, and mentoring. These supports help neurodivergent learners develop self-advocacy skills, build social networks, and navigate the challenges of independent living.

Assistive Technology

Assistive technology (AT) refers to tools, devices, software, or equipment that support individuals with disabilities in performing tasks, enhancing their independence, and facilitating access to educational settings and materials. For neurodivergent learners, AT can be an invaluable resource for improving communication, organization, information processing, and academic skills.

Speech-to-text and text-to-speech software, for example, can assist individuals with dyslexia or language processing difficulties in accessing written information and expressing their thoughts. Visual aids and graphic organizers can help individuals with executive functioning challenges organize their thoughts, plan their tasks, and understand complex concepts.

Other examples of assistive technology include specialized computer programs, adaptive devices, sensory supports, and communication apps. These tools can be tailored to the specific needs of neurodivergent learners, providing them with alternative means of engagement, expression, and participation in educational activities.

Community Partnerships and Wraparound Services

In addition to specialized programs within educational settings, community partnerships and wraparound services provide vital support for neurodivergent learners and their families. These services focus on promoting connectivity, well-being, and inclusive participation in community activities.

Community partnerships involve collaboration between schools, families, service providers, and local organizations to coordinate a range of supports. These may include after-school programs, sports clubs, social groups, and recreational activities that cater to the interests and needs of neurodivergent learners. By engaging in community activities, neurodivergent learners have opportunities to develop social skills, build friendships, and foster a sense of belonging.

Wraparound services, on the other hand, involve the coordination and integration of multiple services and supports to comprehensively address the various needs of neurodivergent individuals and their families. These services may include counseling, therapy, healthcare, respite care, and assistance with daily living skills. The goal is to provide a holistic support system that promotes overall well-being and quality of life.

Real-World Application: The Peers Program

The PEERS (Program for the Education and Enrichment of Relational Skills) program is an evidence-based intervention specifically designed to enhance the social skills and friendship abilities of adolescents and young adults with autism spectrum disorder (ASD). The program targets key social areas, such as conversational skills, cooperation, and problem-solving, and employs a didactic and experiential approach to facilitate skill development.

The PEERS program utilizes a manualized curriculum, and sessions are typically conducted in a group format with trained facilitators. Participants engage in a structured and systematic learning process that includes didactic presentations, interactive activities, role-playing, and real-life practice. The program also involves coaching and feedback to promote generalization of skills in naturalistic social settings.

Research has shown that the PEERS program leads to improvements in social skills, social functioning, and overall quality of social relationships for individuals with ASD. It is frequently implemented in schools, community centers, and clinical settings as an effective intervention for supporting the social development of neurodivergent learners.

Conclusion

Specialized programs and supports are instrumental in creating inclusive educational environments and promoting the success of neurodivergent learners. By tailoring interventions, accommodations, and resources to address the unique needs of neurodivergent individuals, these programs play a crucial role in facilitating academic, social, and emotional growth. Collaboration among educators, families, professionals, and community partners is essential in implementing and sustaining these programs to ensure the well-being and success of neurodivergent learners.

Building Self-Esteem and Confidence in the Classroom

Building self-esteem and confidence in the classroom is essential for promoting the overall well-being and academic success of neurodivergent students. When students feel secure and valued, they are more likely to engage in learning and take risks, leading to improved learning outcomes. In this section, we will explore strategies and approaches that educators can employ to foster self-esteem and confidence in neurodivergent students.

Creating a Positive Classroom Environment

The first step in building self-esteem and confidence is creating a positive and inclusive classroom environment. This can be achieved through several strategies:

1. **Establishing clear expectations:** Providing clear guidelines and expectations for behavior and academic performance helps students feel secure and understand what is expected of them. Communicating these expectations in a positive and supportive manner can enhance students' self-belief and confidence.

2. **Promoting respect and acceptance:** Foster an environment where all students, regardless of their neurodivergent traits, are respected and accepted. Encourage students to embrace differences and appreciate the unique strengths and perspectives that neurodivergent students bring to the classroom.

3. **Encouraging collaboration and peer support:** Facilitate collaborative activities and encourage peer support to create a sense of belonging and camaraderie in the classroom. Pairing neurodivergent students with neurotypical peers can help in building friendships and promote a positive self-image.

4. **Providing a safe space for expression:** Create opportunities for students to express themselves creatively and share their thoughts and ideas without judgment. Allowing students to contribute to classroom discussions and valuing their contributions can boost their self-esteem.

Recognizing and Celebrating Individual Strengths

Every neurodivergent student has unique strengths and talents that should be recognized and celebrated. By focusing on these strengths, educators can help build self-esteem and confidence. Here are some strategies to achieve this:

1. **Strength-based assessments:** Incorporate strength-based assessments that highlight the talents and abilities of neurodivergent students. This approach shifts the focus from deficits to strengths and empowers students by showcasing their abilities.

2. **Opportunities for success:** Provide opportunities for neurodivergent students to succeed and experience a sense of accomplishment. Break down complex tasks into manageable steps, provide additional support when needed, and celebrate their achievements publicly.

3. **Individualized learning plans:** Develop individualized learning plans that take into account each student's strengths and preferences. Tailor teaching approaches and materials to accommodate different learning styles and provide ample opportunities for neurodivergent students to showcase their talents.

4. **Incorporating student interests:** Integrate the interests and hobbies of neurodivergent students into the curriculum whenever possible. This not only enhances their engagement in learning but also reinforces their sense of self-worth and confidence.

Encouraging Growth Mindset

A growth mindset is the belief that abilities and intelligence can be developed through effort, practice, and persistence. By promoting a growth mindset, educators can help neurodivergent students develop resilience, overcome challenges, and cultivate self-esteem. Here's how to nurture a growth mindset in the classroom:

1. **Framing failure as learning opportunities:** Emphasize that making mistakes and facing setbacks are valuable learning experiences. Encourage neurodivergent students to reframe failures as opportunities to grow and improve rather than sources of frustration or disappointment.

2. **Providing constructive feedback:** Offer specific and constructive feedback that focuses on effort, strategies, and improvement rather than on fixed abilities or outcomes. Help students identify areas for growth and provide guidance on how to develop their skills further.

3. **Teaching problem-solving and resilience:** Teach problem-solving skills and resilience-building strategies explicitly. Encourage students to persevere through challenges, break down problems into manageable steps, and seek alternative strategies when facing difficulties.

4. **Modeling a growth mindset:** Model a growth mindset yourself by sharing personal stories of overcoming challenges and working towards personal growth. By showing your own resilience and determination, you inspire neurodivergent students to believe in their ability to improve and succeed.

Providing Emotional Support

Emotional support plays a vital role in building self-esteem and confidence. Here are some strategies to provide emotional support to neurodivergent students:

1. **Creating a nurturing and non-judgmental space:** Foster an emotionally safe and supportive environment where students feel comfortable expressing their emotions and seeking guidance when needed. Establish trust and confidentiality so that students feel secure discussing their concerns.

2. **Developing self-regulation skills:** Teach self-regulation techniques such as deep breathing exercises, mindfulness, and positive self-talk. These practices help neurodivergent students manage stress, anxiety, and other emotions, thereby boosting their self-esteem and confidence.

3. **Individual counseling or support sessions:** Offer individual counseling or support sessions to address the specific emotional needs of neurodivergent students. Provide a listening ear, validate their experiences, and offer guidance on navigating challenges.

4. **Collaborating with parents and caregivers:** Maintain open lines of communication with parents and caregivers to ensure a holistic approach to supporting neurodivergent students. Share strategies, discuss concerns, and develop joint interventions that promote emotional well-being.

Remember, building self-esteem and confidence is an ongoing process. It requires consistent effort, empathy, and understanding from educators. By implementing these strategies, educators can create a nurturing and inclusive classroom environment that fosters the growth and development of neurodivergent students.

The Role of Assistive Technology in Learning

Assistive technology (AT) plays a crucial role in supporting the learning needs of neurodivergent individuals. AT refers to any device, equipment, or software that helps individuals with disabilities overcome barriers to learning and enhance their educational experience. In the context of neurodiversity, AT can assist individuals with diverse learning profiles, such as autism spectrum disorder (ASD), attention-deficit/hyperactivity disorder (ADHD), dyslexia, and intellectual disabilities, in maximizing their potential and achieving academic success.

1. Understanding the Needs of Neurodivergent Learners

Neurodivergent learners often have unique challenges in the learning environment, which may include difficulties with attention, organization, memory, reading, writing, and communication. AT can address these challenges by providing personalized solutions that cater to their specific needs. It is important to assess the individual's strengths and limitations to determine the most appropriate AT supports.

2. Types of Assistive Technology

There is a wide range of AT tools and strategies available to support neurodivergent learners. These include:

2.1. Text-to-Speech Technology: Text-to-speech (TTS) technology converts written text into spoken words, allowing individuals with reading difficulties or print disabilities to access and comprehend written content. TTS software can be installed on computers, tablets, or smartphones, enabling learners to listen to digital textbooks, articles, and other written materials.

2.2. Speech Recognition Software: Speech recognition software allows individuals to dictate their thoughts and ideas, which are then converted into written text. This tool is beneficial for learners with difficulties in writing or organizational skills. It can enhance their ability to express themselves effectively and independently.

2.3. Graphic Organizers and Mind Mapping Tools: These visual tools help individuals with planning, organizing, and understanding complex information. Mind mapping software allows learners to create diagrams and hierarchical structures to organize their thoughts visually. This aids in comprehension, memory, and information recall.

2.4. Assistive Reading and Writing Software: This software provides features such as word prediction, spell-check, and grammar-check, aiding learners with dyslexia, ADHD, or writing difficulties. It assists in improving writing skills, reducing errors, and enhancing overall written communication.

2.5. Time Management and Organization Tools: These tools, including digital calendars, task managers, and reminders, help neurodivergent learners stay organized, manage their time effectively, and meet deadlines. They can also provide visual schedules and checklists to aid in task completion and enhance executive functioning skills.

3. Considerations for Implementing Assistive Technology

While AT can greatly benefit neurodivergent learners, its successful implementation requires careful consideration of various factors:

3.1. Individualized Approach: AT solutions should be tailored to the unique needs and preferences of each learner. It is essential to involve the learner, their teachers, and caregivers in the decision-making process to ensure that the chosen AT tools align with their specific requirements.

3.2. Accessibility and Usability: AT tools should be user-friendly and accessible across multiple devices or platforms to facilitate seamless integration into the learning environment. Compatibility with various operating systems and assistive technology standards is crucial to ensure widespread usability.

3.3. Training and Support: Proper training and ongoing support are vital for the effective use of AT. Educators, parents, and learners should receive training on how to use the specific AT tools and understand their features and functionalities. Technical support should also be readily available to address any issues or challenges that may arise during implementation.

3.4. Assistive Technology Evaluation: Regular evaluation of the effectiveness of AT tools is essential to ensure that they continue to meet the evolving needs of the learners. This evaluation may involve gathering feedback from the learners, monitoring progress, and making necessary adjustments or changes as required.

4. Success Stories and Real-World Examples

Many neurodivergent individuals have benefited from the use of AT in their educational journeys. For example, a student with dyslexia may use text-to-speech software to access written materials, improving their reading comprehension and overall academic performance. Another student with ADHD may use time management and organization tools to stay focused, manage assignments effectively, and meet deadlines. These success stories highlight the transformative power of AT in supporting neurodivergent learners.

5. Additional Resources and Caveats

To explore the world of assistive technology further, numerous resources are available. Educators, parents, and learners can access online platforms, webinars, conferences, and workshops on AT implementation and best practices. It is important to note that AT should not replace effective teaching practices or accommodations within the learning environment. Instead, it should complement and enhance existing strategies to provide comprehensive support for neurodivergent learners.

In conclusion, assistive technology plays a pivotal role in supporting neurodivergent learners' diverse needs in the educational setting. By providing personalized solutions, AT helps overcome barriers to learning and allows individuals to reach their full potential. The successful implementation of AT requires a thoughtful and individualized approach, ensuring accessibility, usability, and ongoing support. With the right AT tools and strategies in place, neurodivergent learners can thrive and succeed in their educational journeys.

Transitioning to Higher Education for Neurodivergent Students

Transitioning to higher education can be a challenging and exciting time for all students, but it can be particularly overwhelming for neurodivergent students.

Neurodivergent students may face unique barriers and require additional support to navigate the academic and social aspects of college or university life. This section will explore the specific challenges faced by neurodivergent students and provide strategies and resources to facilitate a successful transition to higher education.

Understanding the Challenges

Neurodivergent students, such as those with autism, ADHD, dyslexia, or intellectual disabilities, may encounter various difficulties when transitioning to higher education. These challenges can arise in different areas, including academics, social interactions, executive functioning, and self-advocacy. It is crucial to understand these challenges to effectively support neurodivergent students during their transition.

Academic Challenges Neurodivergent students may experience academic challenges due to their unique learning profiles. For example, students with dyslexia may struggle with reading and written assignments, while those with ADHD may have difficulty with attention, organization, and time management. Understanding these challenges is essential for developing appropriate accommodations and support strategies.

Social Interactions Neurodivergent students may have difficulties navigating social interactions in higher education settings. They may struggle with interpreting social cues, making friends, and engaging in group activities. The stress and anxiety associated with social situations can impact their overall well-being and academic success. It is crucial to provide resources and support to help neurodivergent students develop social skills and build social connections.

Executive Functioning Executive functioning skills, such as planning, organization, time management, and problem-solving, are essential for academic success. Neurodivergent students may face challenges in these areas, making it harder for them to stay organized, prioritize tasks, and manage their time effectively. Teaching and reinforcing executive functioning skills can greatly benefit neurodivergent students during their transition to higher education.

Self-Advocacy Self-advocacy is an important skill for all students, but it is especially crucial for neurodivergent students. They need to be able to communicate their needs, seek support, and navigate the academic and social systems of the university or college. Building self-advocacy skills early on can empower neurodivergent students to access the necessary accommodations and resources.

Strategies for a Successful Transition

Transitioning to higher education can be made smoother for neurodivergent students with the implementation of appropriate strategies and supports. Here are some strategies that can facilitate a successful transition:

Preparation and Transition Programs Providing neurodivergent students with pre-transition programs can help ease their transition to higher education. These programs can offer orientation sessions, campus tours, and opportunities to meet support staff and other neurodivergent students. These initiatives help neurodivergent students become familiar with the campus, understand the academic expectations, and build a support network before their official start.

Individualized Education Plan (IEP) and Accommodations Transition planning should start well before entering higher education. Neurodivergent students who have Individualized Education Plans (IEPs) or similar accommodations in high school should collaborate with their support team to create a transition plan. This plan can outline necessary accommodations and supports that will help them succeed academically and socially in higher education. These accommodations may include extra time on exams, note-taking assistance, or access to assistive technology.

Access to Support Services Higher education institutions should have dedicated support services for neurodivergent students. These services may include disability offices, counseling centers, academic coaches, and peer mentoring programs. These support services can provide guidance, counseling, and academic support tailored to the unique needs of neurodivergent students. It is important for neurodivergent students to be aware of and utilize these resources to their fullest potential.

Faculty and Staff Training Training faculty and staff on neurodiversity and the specific needs of neurodivergent students is crucial for creating an inclusive educational environment. By providing training and awareness sessions, higher education institutions can foster a more supportive and understanding community. Faculty and staff members who are knowledgeable about neurodiversity can better accommodate the diverse learning styles and needs of neurodivergent students.

Peer Support Networks Establishing peer support networks for neurodivergent students can greatly enhance their college experience. Connecting incoming neurodivergent students with current students who have successfully transitioned to higher education can provide valuable insights, guidance, and a supportive community. Peer mentors can offer advice on navigating academic courses, managing workload, and participating in extracurricular activities.

Building Self-Advocacy Skills Empowering neurodivergent students to develop self-advocacy skills is essential for their success in higher education. Institutions should provide training and resources that help neurodivergent students understand their rights, communicate their needs effectively, and navigate university policies and procedures. Additionally, mentoring programs and workshops can offer guidance on self-advocacy strategies and techniques.

Real-World Examples and Resources

There are many resources available to assist neurodivergent students during their transition to higher education. Here are a few examples:

Autism on Campus Autism on Campus is an organization that aims to improve the college experience for autistic students. They provide resources, mentorship programs, and a supportive online community for autistic students transitioning to higher education.

Understood Understood is a comprehensive website that offers resources and support for parents, educators, and individuals with learning and attention issues. Their resources include information on the transition to college or university for neurodivergent students.

Student Accessibility Services Most higher education institutions have Student Accessibility Services or a similar department that offers support for students with disabilities. These services provide accommodations, resources, and support tailored to the individual needs of neurodivergent students.

Technology Tools Various technology tools can assist neurodivergent students in higher education. For example, text-to-speech software can help students with reading difficulties, while organizational apps can support executive functioning skills. Exploring available assistive technology options can greatly benefit neurodivergent students in their academic pursuits.

Real-World Example: College Transition Program ABC University offers a College Transition Program for neurodivergent students. The program provides a week-long orientation where students can familiarize themselves with the campus, meet key faculty and staff, and learn about available resources. Throughout the academic year, students receive ongoing support from mentors and have access to a dedicated support office.

Exercises

1. Identify three potential academic challenges that a dyslexic student might face when transitioning to higher education. Provide strategies to help address each challenge.

2. Discuss the importance of faculty and staff training on neurodiversity in creating an inclusive educational environment. Provide examples of training sessions that could be beneficial for faculty and staff.

3. Research and list three assistive technology tools that can support executive functioning skills in neurodivergent students. Explain how each tool can be beneficial.

4. Imagine you are a peer mentor for a neurodivergent student transitioning to higher education. Create a step-by-step guide outlining the support and resources you would provide to help them navigate their first semester successfully.

5. Reflect on a personal experience or a case study of a neurodivergent student's transition to higher education. Discuss the challenges faced and the strategies that were effective in facilitating a successful transition.

Conclusion

Transitioning to higher education can be a transformative and challenging experience for neurodivergent students. By understanding the unique challenges they face and implementing appropriate strategies and supports, higher education institutions can create an inclusive and empowering environment for neurodivergent students. With the right resources and guidance, neurodivergent students can thrive academically, socially, and personally during their college or university journey.

Teacher Training and Professional Development

Promoting Neurodiversity Competence in Teachers

One of the key factors in creating an inclusive and accepting educational environment for neurodivergent students is promoting neurodiversity competence among teachers. Neurodiversity competence refers to a teacher's knowledge, understanding, and ability to effectively support and accommodate students with diverse neurodevelopmental conditions. By increasing their neurodiversity competence, teachers can create a more inclusive classroom environment, foster positive learning experiences, and help neurodivergent students reach their full potential.

To promote neurodiversity competence among teachers, several strategies can be implemented:

Professional Development and Training

Providing teachers with professional development opportunities focused on neurodiversity is crucial in fostering competence. These training programs should cover a wide range of topics, including an overview of neurodevelopmental conditions, the principles of the neurodiversity paradigm, effective teaching strategies, and accommodations for students with diverse needs. It is essential to engage teachers in interactive sessions that incorporate hands-on activities, case studies, and real-life examples to facilitate active learning and application of knowledge.

Collaboration with experts in the field, such as neurodiversity advocates, neurologists, psychologists, and special education professionals, can enhance the effectiveness of professional development programs. These experts can provide insights, share best practices, and answer teachers' questions, ensuring they receive accurate and up-to-date information.

Building Awareness and Understanding

To promote neurodiversity competence, it is crucial to help teachers develop a deep understanding of neurodivergent individuals and their experiences. This can be achieved by sharing personal narratives and stories of individuals with neurodivergent traits, providing firsthand accounts of their challenges, strengths, and unique perspectives. Hearing these stories can help teachers recognize and appreciate the diversity of neurodivergent students and challenge preconceived notions or stereotypes they may hold.

Additionally, teachers should be encouraged to engage in self-reflection and examine their own attitudes, biases, and assumptions related to neurodivergent individuals. This process of self-reflection helps create a mindset shift towards embracing neurodiversity and moving away from deficit-based thinking.

Creating Inclusive Classroom Environments

An inclusive classroom environment is essential for supporting neurodivergent students. Teachers can promote neurodiversity competence by implementing universal design for learning (UDL) principles in their instructional practices. UDL focuses on designing lessons and materials that are accessible to all students, regardless of their neurodevelopmental profiles. This approach involves offering multiple means of representation, engagement, and expression to cater to diverse learning styles and strengths.

Teachers can also prioritize factors like environmental modifications, such as reducing sensory distractions, creating visual supports, and providing flexible seating arrangements, to meet the needs of neurodivergent students. It is important to establish clear routines, communicate expectations effectively, and provide individualized supports to promote a sense of safety and predictability in the classroom.

Collaboration and Communication

To support neurodivergent students effectively, teachers must foster collaboration and open communication with families, support professionals, and other relevant stakeholders. Maintaining a strong partnership with families ensures that teachers have a comprehensive understanding of students' strengths, challenges, and individualized accommodations. Collaborating with support professionals, such as special education teachers, counselors, and speech and language therapists, can also provide valuable insights and strategies for supporting neurodivergent students.

Regular communication with families and professionals should include discussing progress, goal planning, and addressing concerns in a collaborative and solution-oriented manner. Teachers can facilitate this communication by implementing systems such as regular meetings, progress reports, and utilizing technological platforms that promote easy and efficient information sharing.

Continuous Learning and Reflection

Promoting neurodiversity competence among teachers is an ongoing process that requires a commitment to continuous learning and reflection. Teachers should be encouraged to seek out new resources, attend conferences and workshops, and stay updated with current research and practices in the field of neurodiversity. Engaging in personal reflection and self-evaluation allows teachers to identify areas for growth and implement strategies to improve their practice continually.

Teachers can establish professional learning communities or networks to engage in discussions, share experiences, and support one another in promoting neurodiversity competence. These communities can provide a platform for learning from colleagues, exchanging ideas, and brainstorming solutions to common challenges faced in the classroom.

In conclusion, promoting neurodiversity competence among teachers is crucial for creating inclusive and supportive learning environments for neurodivergent students. By providing professional development opportunities, building awareness and understanding, creating inclusive classrooms, fostering collaboration and communication, and encouraging continuous learning, teachers can enhance their competence and make a positive impact on the educational journeys of neurodivergent students.

Addressing Bias and Stereotypes in Education

Addressing bias and stereotypes in education is crucial for creating inclusive and equitable learning environments for all students. Bias refers to the tendency to favor or hold prejudiced views towards certain groups or individuals, while stereotypes are generalized beliefs or assumptions about a particular group. In the context of education, bias and stereotypes can have harmful effects on neurodivergent students, who often face stigma and discrimination due to their differences.

To address bias and stereotypes effectively, it is essential to raise awareness among educators, administrators, and students about the importance of embracing neurodiversity and challenging preconceived notions about individuals with neurodevelopmental conditions. This can be achieved through various strategies and approaches, including:

1. **Professional development and training:** Educators should receive comprehensive training on neurodiversity, including understanding the strengths and challenges of neurodivergent students, debunking common stereotypes, and learning about evidence-based strategies for supporting their needs in the classroom. This training should also emphasize the promotion of inclusive practices and the use of person-centered approaches.

2. **Cultivating empathy and perspective-taking:** Creating opportunities for students to engage in empathy-building activities can help reduce bias and stereotypes. By encouraging students to put themselves in the shoes of

neurodivergent individuals and understand their experiences, educators can foster a more inclusive and accepting classroom environment.

3. **Curriculum integration:** Incorporating diverse perspectives and experiences into the curriculum can challenge stereotypes and promote understanding. This includes featuring literature, videos, and other educational materials that highlight neurodivergent individuals' achievements and contributions. By doing so, educators can foster an environment that celebrates differences and promotes inclusivity.

4. **Promoting positive role models:** Showcasing neurodivergent individuals who have excelled in various fields can inspire students and challenge stereotypes. Inviting neurodivergent guest speakers or organizing panel discussions can provide students with real-life examples of successful individuals and allow them to engage in meaningful conversations about neurodiversity.

5. **Encouraging open dialogue:** Creating a safe space for conversations about bias and stereotypes is essential. Educators should encourage students to ask questions, share their thoughts, and engage in respectful discussions about neurodiversity. This can help dispel misconceptions, challenge biases, and promote a more inclusive and understanding classroom community.

6. **Addressing language use:** Educators should be mindful of the language they use when referring to neurodivergent students. Avoiding derogatory terms or labels and using person-first language (e.g., referring to students as "students with autism" rather than "autistic students") can foster a more respectful and inclusive environment.

7. **Promoting collaboration and peer support:** Encouraging collaboration among students of diverse backgrounds, including neurodivergent and neurotypical students, can foster empathy, understanding, and acceptance. Implementing peer support programs, such as buddy systems or mentoring initiatives, can further enhance inclusivity and provide additional support for neurodivergent students.

It is important to note that addressing bias and stereotypes should be an ongoing and continuous process. Educators should regularly evaluate their own beliefs and biases, reflect on their teaching practices, and seek feedback from students and colleagues. By actively working to dismantle biases and challenge stereotypes, educators can create an inclusive and supportive learning environment that celebrates the strengths and differences of all students.

Example Problem:

Sarah, a high school teacher, notices that one of her neurodivergent students, David, is often excluded by his peers during group activities. Sarah believes that the other students' behavior may stem from stereotypes and lack of understanding about neurodivergent individuals. How can Sarah address this situation and promote inclusion?

Solution:

To address the exclusion of David and promote inclusion in the classroom, Sarah can implement the following strategies:

1. **Education and awareness:** Sarah can plan a lesson on neurodiversity, providing her students with information about different neurodevelopmental conditions. She can emphasize the strengths and unique perspectives that neurodivergent individuals bring to the classroom. By increasing students' understanding and challenging stereotypes, Sarah can foster a more inclusive environment.

2. **Collaborative learning activities:** Sarah can design cooperative learning tasks that require students to work together and value each other's contributions. By structuring group activities that highlight individual strengths and promote collaboration, Sarah can help her students recognize and appreciate the diverse abilities and talents within the classroom.

3. **Social skills development:** Sarah can plan explicit instruction on social skills, such as active listening, empathy, and effective communication. By teaching and modeling these skills, Sarah can help her students develop a better understanding of each other's perspectives and encourage them to embrace diversity.

4. **Peer support:** Sarah can assign a peer buddy to David, who can offer support and friendship during group activities. Having a designated peer buddy can help David feel included and accepted, enhancing his overall classroom experience.

5. **Classroom culture:** Sarah can create a classroom culture that celebrates diversity and promotes inclusivity. This can be achieved through the use of inclusive language, displays of diverse role models, and encouraging open dialogue about differences.

By implementing these strategies, Sarah can create a more inclusive and accepting classroom environment, where all students, including David, feel valued and supported.

Creating Inclusive Classroom Environments

Creating inclusive classroom environments is essential for supporting neurodivergent students and promoting their success. In an inclusive classroom, all students, regardless of their neurodiversity, feel welcome, valued, and supported in their learning. This section will discuss the principles, strategies, and best practices for creating inclusive classrooms that foster the academic, social, and emotional growth of neurodivergent students.

Understanding Inclusion

Inclusion in the classroom refers to the practice of accommodating diverse learning needs and ensuring equitable access to education for all students. It goes beyond simply integration, where neurodivergent students are placed in mainstream classrooms without adequate support. Inclusive classrooms embrace the idea that all students, including those with neurodivergent conditions, have unique strengths, abilities, and perspectives that contribute to the learning community.

Universal Design for Learning (UDL)

One of the key frameworks for creating inclusive classroom environments is Universal Design for Learning (UDL). UDL is an educational approach that provides flexibility in how students access, engage with, and demonstrate their learning. It recognizes that learners have different preferences, strengths, and challenges, and aims to design learning experiences that cater to these individual differences.

The three principles of UDL include:

1. Multiple means of representation: Providing information and content in multiple formats to accommodate diverse learning styles and preferences. This can include visual aids, audio recordings, videos, and hands-on activities.

2. Multiple means of engagement: Offering various ways for students to engage with the learning material, fostering motivation and interest. This can involve providing choices, scaffolding, and incorporating real-world connections.

3. Multiple means of action and expression: Allowing students to demonstrate their knowledge and understanding through different modes of communication and expression. This can include written assignments, oral presentations, artistic creations, and multimedia projects.

By applying UDL principles, educators can create learning environments that meet the needs of all students, including those with learning differences or neurodivergent conditions.

Individualized Education Plans (IEPs) and Supports

For neurodivergent students who require additional support, Individualized Education Plans (IEPs) play a crucial role in ensuring their success in inclusive classrooms. An IEP is a personalized document that outlines specific goals, accommodations, and supports tailored to the student's unique needs and learning objectives.

IEPs are developed collaboratively with input from parents or caregivers, educators, and other professionals involved in the student's education. They provide a roadmap for educators to implement strategies, modifications, and accommodations that will enable the student to access the curriculum, participate in classroom activities, and demonstrate their learning.

Examples of supports and accommodations that may be included in an IEP for neurodivergent students include:

- Extra time for completing assignments or tests - Preferential seating to optimize focus and minimize distractions - Assistive technology tools such as text-to-speech software or touch-screen devices - Visual supports like visual schedules, graphic organizers, or visual timers - Use of fidget tools or sensory breaks to manage overstimulation or promote focus - Peer support or buddy systems to facilitate social interactions and inclusion - Specific instructional strategies tailored to the student's learning style or preferences

IEPs should be regularly reviewed and updated to reflect the student's progress, evolving needs, and any changes in the learning environment.

Differentiated Instruction and Teaching Strategies

In an inclusive classroom, educators employ differentiated instruction strategies to meet the diverse needs of neurodivergent students. Differentiated instruction involves adapting the content, process, and learning environment to accommodate varying learning styles, abilities, and interests.

Some strategies for differentiated instruction in inclusive classrooms include:

- Varied instructional materials: Providing a range of materials, texts, and resources to cater to different reading levels, learning preferences, and interests.

- Flexible grouping: Organizing students into small groups based on their learning needs and facilitating cooperative learning experiences. This allows for peer collaboration, support, and personalized instruction.

- Scaffolded instruction: Providing additional support and guidance during the learning process. This can involve breaking down complex tasks into smaller, more manageable steps or gradually releasing responsibility to students.

- Choice boards: Offering students a range of tasks or activities, allowing them to select one that aligns with their interests or strengths. This fosters engagement, motivation, and ownership over their learning.

- Multi-modal approaches: Incorporating a variety of instructional techniques and materials, such as visual aids, manipulatives, technology, real-life examples, and hands-on activities. This accommodates diverse learning preferences and enhances understanding and retention.

By differentiating instruction, educators can create an inclusive learning environment where all students can actively participate, engage, and achieve their full potential.

Collaborative and Multidisciplinary Approaches in Education

Collaboration and multidisciplinary approaches are essential for creating inclusive classroom environments. Educators, specialists, support staff, and families must work together to address the diverse needs of neurodivergent students and ensure their holistic development.

Collaboration can take various forms, including:

- Regular team meetings: Bringing together educators, specialists, and support staff to discuss student progress, share insights, coordinate interventions, and develop strategies.

- Individualized supports: Providing interventions, therapies, or services that are tailored to the individual needs of students. This may involve collaboration with speech-language pathologists, occupational therapists, or behavioral specialists.

- Family involvement: Engaging parents and caregivers as partners in the education process. This can include regular communication, parent-teacher conferences, sharing resources, and involving families in decision-making regarding their child's education.

- Professional development: Offering opportunities for educators to enhance their knowledge and skills in supporting neurodivergent students. This can involve workshops, seminars, peer observations, and ongoing training.

Through collaboration and multidisciplinary approaches, educators can tap into a diverse range of expertise, perspectives, and resources to create truly inclusive classroom environments.

Promoting Positive Classroom Culture

Creating an inclusive classroom goes beyond implementing strategies and accommodations. It necessitates fostering a positive classroom culture that celebrates diversity, promotes acceptance, and cultivates empathy and respect among students.

Some strategies for promoting a positive classroom culture include:

- Curriculum integration: Incorporating diverse perspectives, cultures, and experiences in the curriculum to promote understanding, empathy, and inclusion.

- Social-emotional learning (SEL): Integrating SEL programs and practices that foster self-awareness, self-regulation, social skills, and responsible decision-making. This helps students develop empathy and understanding for their neurodivergent peers.

- Peer support programs: Establishing peer support programs or buddy systems where neurotypical students can provide support and mentorship to their neurodivergent classmates. This promotes social inclusion and positive relationships among students.

- Classroom discussions and reflections: Facilitating open and respectful discussions about neurodiversity, promoting understanding, dispelling myths, and addressing any stereotypes or misconceptions.

- Anti-bullying initiatives: Implementing comprehensive anti-bullying programs to create a safe and nurturing environment for all students. This involves setting clear expectations, providing education on diversity and empathy, and promptly addressing any incidents of bullying or discrimination.

By prioritizing a positive classroom culture, educators can create an inclusive learning environment where all students feel valued, respected and can thrive academically, socially, and emotionally.

Key Takeaways

- Inclusive classrooms aim to accommodate diverse learning needs and ensure equitable access to education for all students.

- Universal Design for Learning (UDL) provides a framework for accommodating diverse learning needs by offering multiple means of representation, engagement, and action and expression.

- Individualized Education Plans (IEPs) are personalized documents that outline goals, accommodations, and supports for neurodivergent students.

- Differentiated instruction involves adapting content, process, and learning environment to cater to different learning styles, abilities, and interests.

- Collaboration and multidisciplinary approaches are crucial for meeting the diverse needs of neurodivergent students and involve regular team meetings, individualized supports, family involvement, and professional development.

- Promoting a positive classroom culture involves integrating diverse perspectives, fostering social-emotional learning, establishing peer support programs, facilitating open discussions, and implementing anti-bullying initiatives.

Collaboration with Families and Support Networks

Collaboration with families and support networks is a crucial aspect of promoting the well-being and success of neurodivergent individuals. These stakeholders play a vital role in creating a supportive environment at home and in the community, and their involvement can greatly contribute to the overall development and advancement of neurodivergent individuals. This section explores the importance of collaboration, strategies for effective collaboration, and the resources available to families and support networks.

Importance of Collaboration

Collaboration between professionals, families, and support networks is essential for the holistic support of neurodivergent individuals. By working together, we can create a comprehensive plan that addresses the unique needs and challenges of each individual. Collaboration ensures that everyone involved has a common understanding of the goals and objectives, and can pool their expertise and resources to provide the best possible support.

Families and support networks possess invaluable knowledge about the strengths, preferences, and interests of the neurodivergent individual. This knowledge is essential for designing effective interventions and accommodations. Collaboration also helps professionals gain insights into the daily experiences and challenges faced by the neurodivergent individual, enabling them to tailor their support accordingly.

Furthermore, collaboration fosters a sense of partnership, trust, and respect among all stakeholders. It empowers families and support networks to actively participate in decision-making processes, leading to more personalized and meaningful interventions. Additionally, collaboration promotes continuity of care and consistency in support, as information and strategies can be shared and coordinated across different contexts and settings.

Strategies for Effective Collaboration

Effective collaboration requires open communication, mutual respect, and a shared commitment to the well-being and development of the neurodivergent individual. Here are some strategies to facilitate collaboration with families and support networks:

1. **Establishing a Partnership:** Begin by recognizing families and support networks as vital partners in the individual's journey. Value their perspectives,

expertise, and experiences. Foster open and transparent communication channels to ensure that everyone is involved in decision-making processes.

2. **Sharing Information:** Provide families and support networks with comprehensive information about the neurodivergent individual's condition, strengths, challenges, and available support services. This enables them to make informed decisions and actively contribute to interventions and accommodations. Use clear and jargon-free language, and encourage questions and discussions to address any concerns or misconceptions.

3. **Active Listening:** Listening actively and empathetically to the concerns and perspectives of families and support networks is key to effective collaboration. Create a safe and non-judgmental space where they feel comfortable sharing their experiences, challenges, and aspirations. Validate their feelings and experiences, and make them an active part of the problem-solving process.

4. **Individualized Planning:** Collaborate with families and support networks to develop individualized plans that align with the neurodivergent individual's goals, strengths, and priorities. Involve them in goal-setting, decision-making, and monitoring progress. Regularly review and adjust plans as needed, incorporating their feedback and insights.

5. **Building Support Networks:** Facilitate the creation of support networks for families and support networks themselves. Connect them with other families who share similar experiences, as well as relevant support groups and organizations. This allows them to access emotional support, share knowledge and resources, and learn from each other's experiences.

Resources for Families and Support Networks

Numerous resources are available to support families and support networks in their journey of collaboration and support. Here are some key resources to consider:

1. **Education and Training Programs:** Many organizations offer educational programs and workshops that provide families and support networks with essential knowledge, skills, and strategies. These programs cover various topics, such as understanding neurodiversity, effective communication, behavior management, and accessing community resources.

2. **Support Groups and Online Communities:** Support groups and online communities bring together families and support networks to share experiences, provide emotional support, and exchange knowledge and resources. These platforms provide a safe and inclusive space for networking, learning, and building relationships with others who understand their journey.

3. Family-Centered Services: Seek out family-centered services that prioritize the involvement and empowerment of families and support networks. These services recognize the expertise of families and ensure that their voices are heard in decision-making processes. They promote collaboration and partnership, providing families with guidance, advocacy support, and access to various resources.

4. Advocacy Organizations: Many advocacy organizations specialize in providing support and resources for families and support networks. These organizations offer information, guidance, and advocacy services to ensure that the voices of neurodivergent individuals and their families are heard at various levels, including policy-making and service provision.

5. Online Information and Toolkits: Numerous websites and online platforms provide valuable information, toolkits, and resources for families and support networks. These resources cover a wide range of topics, including education, employment, mental health, and self-advocacy. They offer practical strategies, research-based information, and real-life examples to support families in their journey of collaboration and support.

Example Scenario

To illustrate the importance of collaboration with families and support networks, let's consider the case of a teenager with autism spectrum disorder (ASD). The individual's family is actively involved in their educational journey and collaborates with the school and support professionals.

By collaborating with the family, the school gains insights into the individual's strengths, challenges, and interests. The family shares valuable information about the individual's sensory sensitivities, communication preferences, and specific learning styles. Armed with this knowledge, the school can tailor their teaching strategies, provide necessary accommodations, and create a supportive environment that promotes the individual's inclusion and success.

The family, on the other hand, benefits from the collaboration by gaining a deeper understanding of the school system, educational practices, and available support services. They can actively participate in the development of the individualized education plan (IEP), contribute to goal-setting, and advocate for their child's needs. The collaboration between the family and the school ensures that all stakeholders work together to create a supportive and inclusive educational experience for the neurodivergent individual.

Conclusion

Collaboration with families and support networks is vital for the well-being and success of neurodivergent individuals. It promotes a sense of partnership, facilitates the exchange of knowledge and resources, and ensures that interventions and accommodations are personalized and meaningful. By valuing and involving families and support networks as active partners, we can create an environment that celebrates neurodiversity and fosters the holistic development of neurodivergent individuals.

Best Practices for Supporting Neurodivergent Students

Supporting neurodivergent students in an educational setting requires a multifaceted approach that takes into account their unique strengths and challenges. In this section, we will explore some of the best practices for creating an inclusive learning environment that supports the diverse needs of neurodivergent students.

Understanding Neurodivergent Students

To effectively support neurodivergent students, it is crucial to have a deep understanding of their specific neurodevelopmental conditions, such as Autism Spectrum Disorder (ASD), Attention-Deficit/Hyperactivity Disorder (ADHD), and dyslexia. Educators should familiarize themselves with the characteristics, strengths, and challenges associated with each condition, as well as the individual needs of their neurodivergent students.

By having a comprehensive understanding of neurodivergent conditions, educators can better tailor their instructional strategies to meet the diverse learning styles and preferences of their students. This includes recognizing that neurodivergent students may have different ways of processing information, communicating, and interacting with their peers.

Creating a Neurodiverse-Friendly Classroom

The physical environment of the classroom plays a significant role in supporting neurodivergent students. Here are some best practices for creating a neurodiverse-friendly learning space:

- Provide a quiet area: Neurodivergent students may be more sensitive to sensory stimuli, such as noise and fluorescent lighting. Designating a quiet area in the classroom where students can go when they need a break or to reduce sensory overload can be immensely helpful.

- Utilize visual aids: Visual supports, such as visual schedules, visual timers, and visual cues, can enhance understanding and provide structure for neurodivergent students. Visual aids can help with comprehension, organization, and time management.

- Establish clear routines and expectations: Consistent routines and clear expectations can provide a sense of predictability and reduce anxiety for neurodivergent students. Clearly communicate daily schedules, assignments, and behavioral expectations, and provide visual reminders as needed.

- Incorporate flexible seating options: Offering flexible seating choices, such as stability balls, standing desks, bean bags, or fidget tools, can accommodate the sensory needs and preferences of neurodivergent students. This can also promote active engagement and help manage restlessness.

- Designate a safe space: Neurodivergent students may experience heightened social or sensory difficulties in certain situations. Creating a designated safe space in the classroom where students can retreat to when they need a break or support can foster a sense of security and well-being.

Differentiated Instruction and Universal Design for Learning

Differentiated instruction is a key component of supporting neurodivergent students. It involves modifying teaching methods and materials to meet the individual needs and interests of learners with diverse abilities and learning styles. Some strategies for differentiated instruction include:

- Provide multiple means of representation: Present information in different formats to accommodate various learning styles. For example, use visual aids, auditory explanations, hands-on activities, and technology-based resources to make content accessible to all students.

- Offer multiple means of expression: Allow students to demonstrate their understanding and mastery of content through different modes of expression. This can include written assignments, oral presentations, artistic projects, or multimedia creations. Providing choices empowers students and promotes engagement.

- Provide multiple means of engagement: Create opportunities for active participation and personalization in the learning process. Allow students to connect their own experiences and interests to the curriculum. Incorporate hands-on activities, group projects, and collaborative learning experiences.

Universal Design for Learning (UDL) is another framework that promotes inclusive education for all students, including those with neurodivergent conditions. UDL is based on three principles:

- Multiple means of representation: Provide information in different formats and modalities to cater to diverse learner needs. This can include text, images, videos, and interactive materials.
- Multiple means of action and expression: Offer alternative ways for students to demonstrate their knowledge and skills. This can involve options for written or verbal response, physical demonstrations, or use of assistive technology.
- Multiple means of engagement: Foster student engagement by providing choices, personalized learning opportunities, and varied instructional strategies. This includes incorporating real-world contexts, promoting collaboration, and offering supportive feedback and reflection.

By implementing differentiated instruction and UDL principles, educators can ensure that neurodivergent students have equitable access to the curriculum and are actively engaged in their learning.

Building Relationships and Collaboration

Building positive relationships with neurodivergent students is essential for their academic, social, and emotional growth. Here are some strategies for fostering strong relationships and collaboration:

- Practice empathy and understanding: Take the time to understand each student's unique strengths, challenges, and interests. Show empathy and appreciation for their individuality. Create a welcoming and accepting classroom culture that celebrates neurodiversity.
- Foster peer relationships: Encourage opportunities for neurodivergent students to connect with their peers. This can include structured group work, partner activities, or cooperative learning experiences. Provide guidance and support to facilitate positive interactions and friendships.

- Collaborate with support professionals: Work closely with support professionals, such as special education teachers, speech therapists, occupational therapists, and counselors, to develop individualized plans and interventions for neurodivergent students. Regularly communicate and collaborate to ensure consistent support both in and out of the classroom.

- Involve families and support networks: Engage families and support networks in the education of neurodivergent students. Collaborate with parents, guardians, and other caregivers to better understand the individual needs and preferences of their child. Provide resources and strategies for extending support at home.

Promoting Self-Advocacy and Autonomy

Promoting self-advocacy skills and autonomy is crucial for neurodivergent students to become confident, independent learners. Here are some strategies to foster self-advocacy and autonomy:

- Teach self-awareness: Help students develop self-awareness about their strengths, challenges, and learning preferences. Encourage them to reflect on their own learning processes and advocate for their individual needs.

- Build self-advocacy skills: Support students in developing self-advocacy skills, such as expressing their needs, requesting accommodations, or seeking help when necessary. Teach them how to effectively communicate with teachers, peers, and support professionals.

- Foster goal-setting and self-reflection: Guide students in setting meaningful goals and regularly reflect on their progress. Encourage them to take ownership of their learning by planning and monitoring their own academic growth.

- Provide opportunities for choice and autonomy: Offer students choices within assignments, projects, or activities to foster a sense of control and autonomy. This can help build confidence, motivation, and engagement.

- Celebrate strengths and achievements: Recognize and celebrate the unique strengths and achievements of neurodivergent students. This can boost their self-esteem and foster a positive self-concept.

By promoting self-advocacy and autonomy, educators empower neurodivergent students to become active participants in their own education and prepare them for future success.

Professional Development and Continued Learning

Educators should engage in ongoing professional development to stay updated on best practices for supporting neurodivergent students. Professional development opportunities can include workshops, conferences, webinars, and collaboration with other educators. It is important to stay informed about new research, interventions, and evidence-based practices in the field of neurodiversity.

Additionally, educators should actively seek feedback from neurodivergent students, their families, and support professionals to continuously improve their instructional strategies and classroom environment. Reflective practice can help identify areas for growth and enhance the overall educational experience for neurodivergent students.

Furthermore, educators should also consider engaging in self-reflection and self-education to challenge their own biases, assumptions, and preconceptions about neurodivergent individuals. This includes cultivating an open-minded and inclusive mindset that values and celebrates neurodiversity in all aspects of education.

Conclusion

In conclusion, supporting neurodivergent students requires a comprehensive and holistic approach that acknowledges their unique strengths and challenges. By understanding the individual needs of neurodivergent students, creating an inclusive classroom environment, implementing differentiated instruction and UDL principles, building positive relationships, promoting self-advocacy and autonomy, and engaging in professional development, educators can create a supportive and empowering educational experience. By celebrating neurodiversity and embracing the principles of inclusion and acceptance, we can cultivate a neurodiverse-friendly world for the benefit of all.

Employment and Neurodiversity

Neurodivergent Talent in the Workplace

Recognizing and Harnessing Neurodivergent Strengths

Neurodiversity celebrates and values the unique strengths and abilities that individuals with neurodevelopmental conditions possess. While it is important to acknowledge and address the challenges faced by neurodivergent individuals, it is equally crucial to recognize and harness their strengths. By focusing on the positive attributes of neurodivergent individuals, we can build inclusive environments that allow them to thrive and contribute to society in meaningful ways.

Strengths-Based Approach

A strengths-based approach emphasizes the inherent strengths and capabilities of neurodivergent individuals. It is a shift away from a deficit-based perspective that focuses solely on impairments and difficulties. By recognizing and nurturing the strengths of neurodivergent individuals, we can foster their self-confidence, autonomy, and overall well-being.

One key strength often observed in neurodivergent individuals is attention to detail. Their ability to notice and process minute details can lead to exceptional performance in tasks that require precision and accuracy. For example, individuals with autism may excel in fields such as engineering, computer programming, or scientific research, where attention to detail is highly valued. By recognizing and harnessing this strength, organizations can create opportunities for neurodivergent individuals to showcase their skills in these fields.

Another strength commonly found among neurodivergent individuals is the ability to think outside the box and offer unique perspectives. Many individuals with conditions like ADHD and dyslexia have unconventional thinking styles that enable them to approach problems from different angles. This creativity and divergent thinking can be harnessed in fields such as art, design, innovation, and entrepreneurship. By providing neurodivergent individuals with platforms to express their creativity, organizations can benefit from their fresh ideas and unique insights.

Furthermore, neurodivergent individuals often demonstrate exceptional focus and passion for their areas of interest. This intense focus can result in high levels of expertise and achievement in specialized fields. By providing opportunities for neurodivergent individuals to pursue their passions, organizations can leverage their expertise and dedication to make significant contributions in academia, research, and other knowledge-based domains.

Examples of Neurodivergent Strengths

To illustrate the diverse strengths of neurodivergent individuals, let's explore a few examples:

1. Visual Thinking and Spatial Reasoning: People with autism often have strong visual thinking skills and excel in tasks that involve visual-spatial reasoning. This makes them well-suited for careers in architecture, graphic design, and animation, where the ability to think in images and manipulate spatial relationships is essential.

2. Hyperfocus and Task Intensity: Individuals with ADHD may exhibit hyperfocus, a state of intense concentration and engagement in tasks of interest. This strength can be harnessed in professions that require sustained focus and productivity, such as software development, project management, and creative writing.

3. Pattern Recognition and Analytical Thinking: Many individuals with dyslexia have strong pattern recognition abilities and excel in tasks that involve complex problem-solving and analytical thinking. This makes them valuable assets in fields such as data analysis, cryptography, and financial modeling.

4. Empathy and Intuition: Some neurodivergent individuals, particularly those with autism, have a heightened sense of empathy and a unique perspective on the world. This strength can be channeled into careers that require emotional intelligence and a deep understanding of others, such as counseling, social work, and advocacy.

These are just a few examples of the wide range of strengths that neurodivergent individuals possess. Recognizing and harnessing these strengths not only benefits the individuals themselves but also contributes to a more diverse, innovative, and inclusive society.

Creating Opportunities for Neurodivergent Individuals

To fully harness the strengths of neurodivergent individuals, it is important to create opportunities that allow them to thrive. Here are some strategies organizations can employ:

1. Flexible Work Environments: Neurodivergent individuals may thrive in work environments that offer flexibility in scheduling, allowing them to work during their peak performance times. Providing options for remote work or flexible hours can also accommodate their individual needs and preferences.

2. Clear Communication and Expectations: Providing clear expectations and communication channels can support neurodivergent individuals in understanding their roles and responsibilities. Clarifying tasks, providing written instructions, and ensuring open lines of communication can minimize misunderstandings and enhance productivity.

3. Accommodations and Support: Implementing reasonable accommodations, such as noise-cancelling headphones, task organizers, or assistive technologies, can help neurodivergent individuals navigate their work environments more effectively. Offering support resources, like mentoring programs or peer networks, can also foster a sense of belonging and well-being.

4. Training and Education: Providing training and education about neurodiversity to colleagues and supervisors can promote understanding and acceptance. By raising awareness about the unique strengths of neurodivergent individuals, organizations can create an inclusive environment that values and appreciates their contributions.

5. Collaborative and Diverse Teams: Building diverse teams that include neurodivergent individuals can lead to more creative problem-solving and innovative solutions. Creating opportunities for collaboration, where diverse perspectives are valued, can foster a positive work environment that encourages the sharing of ideas and the pooling of strengths.

The Power of Neurodivergent Strengths

Recognizing and harnessing the strengths of neurodivergent individuals not only benefits them as individuals but also has broader societal implications. By embracing neurodiversity in our schools, workplaces, and communities, we can tap into the vast potential and talents of neurodivergent individuals.

When we create environments that value and nurture neurodivergent strengths, we foster inclusion, innovation, and collective well-being. By embracing the diverse ways in which neurodivergent individuals think, learn, and work, we can create a more equitable and accepting society.

The next section will delve into the challenges faced by neurodivergent individuals in the workplace and explore strategies for creating inclusive and supportive environments.

The Concept of Neurodivergent Affinity Spaces

Neurodivergent affinity spaces are community-centered environments that provide a sense of belonging, support, and empowerment for individuals with neurodivergent identities. In these spaces, individuals with diverse neurodivergent conditions, such as autism, ADHD, and dyslexia, come together to connect, share experiences, and celebrate their unique perspectives. These spaces are designed to prioritize the needs and experiences of neurodivergent individuals, fostering a sense of understanding and acceptance.

The concept of neurodivergent affinity spaces is rooted in the principles of neurodiversity, which recognize neurological differences as natural variations of the human brain rather than deficits to be fixed or cured. Affinity spaces provide a platform for neurodivergent individuals to challenge societal norms and perceptions, advocating for acceptance and inclusion in a neurotypical world.

In neurodivergent affinity spaces, participants have the opportunity to engage in a range of activities that promote self-expression, creativity, and personal growth. These may include art workshops, support groups, awareness campaigns, skill-sharing sessions, and social events. The activities are tailored to the unique needs and interests of neurodivergent individuals, promoting a sense of identity and community.

One of the key benefits of neurodivergent affinity spaces is the validation and support they offer. Participants can share their stories, experiences, and challenges without fear of judgment or stigma. This validation helps individuals build self-confidence, develop a positive self-image, and gain a sense of belonging within the broader neurodivergent community. It also encourages self-advocacy, empowering individuals to embrace their neurodivergent identity and advocate for their rights and needs.

In these spaces, individuals with neurodivergent conditions can also access a network of peer support. People with similar experiences and challenges can provide valuable insights, advice, and guidance to one another, fostering a sense of camaraderie and mutual growth. They can navigate shared difficulties, find coping strategies, and celebrate their collective achievements.

Neurodivergent affinity spaces also serve as platforms for education and learning. Participants can engage in workshops and discussions to deepen their understanding of neurodiversity, dismantle stereotypes, and promote acceptance within and beyond the community. By engaging with experts, researchers, and advocates, individuals in these spaces can stay informed about new developments in the field and contribute to ongoing conversations.

It is important to note that neurodivergent affinity spaces are not exclusionary; rather, they aim to create safe and supportive environments that center the experiences of neurodivergent individuals. These spaces welcome allies and advocates who are committed to learning and understanding the nuances of neurodivergent perspectives. By fostering collaboration and shared understanding, affinity spaces contribute to building a neurodiverse society that embraces and celebrates the strengths and unique contributions of all its members.

In conclusion, neurodivergent affinity spaces play a vital role in creating a sense of belonging, empowerment, and community for individuals with neurodivergent identities. These spaces provide a platform for self-expression, validation, support, and education within the broader neurodivergent community. By promoting acceptance, understanding, and allyship, these spaces contribute to reshaping societal attitudes towards neurodiversity and fostering a more inclusive and equitable world for all.

Examples of Successful Neurodivergent Entrepreneurs and Professionals

Neurodivergent individuals have unique talents and abilities that can contribute to their success as entrepreneurs and professionals. In this section, we will explore some inspiring examples of neurodivergent individuals who have achieved significant accomplishments in their respective fields.

1. Temple Grandin

One of the most well-known examples of a successful neurodivergent individual is Temple Grandin. Diagnosed with autism, Grandin overcame significant challenges to become a renowned animal behavior expert and advocate for autism awareness. Her unique perspective, characterized by her visual thinking and attention to detail, enabled her to revolutionize livestock handling systems.

Grandin's designs have been implemented in numerous livestock facilities worldwide, improving animal welfare and the efficiency of the agricultural industry. She has also written several best-selling books, sharing her experiences and insights into autism.

Key Lessons: Grandin's success highlights the importance of embracing neurodivergent strengths and providing opportunities for individuals to excel in their areas of expertise. Her story also emphasizes the significance of recognizing that diverse perspectives can lead to innovative solutions.

2. John Elder Robison

John Elder Robison is another notable example of a successful neurodivergent entrepreneur. Diagnosed with Asperger's syndrome, Robison is recognized for his work as an author, advocate, and pioneer in the field of electronics.

Robison's fascination with electronics began at an early age, leading him to pursue a career in the music industry. He worked as a sound engineer for prominent bands, including KISS and Pink Floyd, and later founded his own successful business, J.E. Robison Service.

As an author, Robison has shared his personal experiences with autism in his memoirs, which have received widespread acclaim. Through his advocacy work, he has played a significant role in raising awareness about autism and promoting neurodiversity acceptance.

Key Lessons: Robison's story demonstrates how embracing individual strengths and passions can pave the way for success. It highlights the importance of providing opportunities for neurodivergent individuals to explore their interests and talents, leading to meaningful contributions in various industries.

3. Hans Asperger

Hans Asperger, the Austrian pediatrician after whom Asperger's syndrome is named, is an intriguing historical example of a successful neurodivergent professional. Asperger exhibited traits of autism himself, although he was not officially diagnosed during his lifetime.

Despite facing personal challenges, Asperger made significant contributions to the understanding and support of individuals with autism. His research and observations formed the basis of what we now know as Asperger's syndrome. He recognized the unique strengths and abilities of his patients and played a pivotal role in promoting inclusive education and acceptance of neurodiversity.

Key Lessons: Asperger's story highlights the potential for individuals to make a lasting impact even in the face of personal challenges. It emphasizes the importance of recognizing and nurturing the talents of neurodivergent individuals to drive positive change in society.

4. Heather Kuzmich

Heather Kuzmich gained recognition through her participation in the reality TV show "America's Next Top Model." Diagnosed with Asperger's syndrome, Kuzmich defied societal expectations and stereotypes by showcasing her unique blend of beauty and neurodivergent traits.

Kuzmich's appearance on the show broke barriers, challenging the fashion industry's narrow definitions of beauty. She used her platform to raise awareness about autism and promote acceptance, becoming an inspiration to many individuals with neurodivergent traits.

Key Lessons: Kuzmich's story illustrates the importance of representation and challenging societal norms. It highlights the value of embracing diversity in all its forms, including neurodivergent individuals, and the potential for positive change through increased visibility.

5. Scott Barry Kaufman

Scott Barry Kaufman is a cognitive psychologist and author who has made significant contributions to the field of intelligence research. Diagnosed with ADHD at a young age, Kaufman encountered numerous challenges throughout his education and career.

Despite these challenges, Kaufman persevered, obtaining a Ph.D. in cognitive psychology and becoming a respected authority in his field. He has written several books on intelligence, creativity, and personal development, encouraging individuals to embrace their unique strengths and pursue their passions.

Key Lessons: Kaufman's story emphasizes the importance of resilience and determination in the face of adversity. It highlights the potential for neurodivergent individuals to excel in academia and make valuable contributions to their respective fields.

These examples of successful neurodivergent entrepreneurs and professionals demonstrate the vast potential and unique perspectives that neurodiversity brings to various industries. They challenge societal stereotypes, promote acceptance, and inspire others to embrace their differences, fostering a more inclusive and diverse world.

The Value of Diversity in Workforce Innovation

In today's rapidly evolving and interconnected world, organizations are recognizing the immense value of diversity in driving innovation and achieving success. This is particularly true when it comes to embracing neurodiversity in the workplace. Neurodiversity refers to the natural variation of neurological functioning in individuals, including conditions such as autism, attention-deficit/hyperactivity disorder (ADHD), dyslexia, and intellectual disabilities. By embracing and accommodating the unique strengths and perspectives of neurodivergent individuals, companies can fuel creativity, problem-solving, and overall workforce innovation.

Benefits of Neurodiversity in the Workplace

Neurodivergent individuals possess a plethora of strengths that can be highly valuable in the workplace. These strengths are often associated with their unique neurodivergent traits and cognitive abilities. For example:

- **Enhanced Attention to Detail:** People with autism and ADHD often excel in focusing on tasks for extended periods, leading to meticulous work and increased accuracy.

- **Pattern Recognition and Analytical Skills:** Individuals with dyslexia and certain types of autism demonstrate exceptional pattern recognition and analytical skills, providing fresh perspectives and innovative problem-solving approaches.

- **Creativity and Original Thinking:** The divergent thinking style often exhibited by neurodivergent individuals results in novel ideas, unconventional solutions, and unique approaches to challenges.

- **Hyperfocus and Intense Dedication:** Neurodivergent individuals can exhibit remarkable levels of passion, dedication, and persistence in pursuing their interests, leading to exceptional expertise and contributions in specific domains.

- **Exceptional Memory and Information Processing:** Many neurodivergent individuals possess exceptional memory and information processing abilities, allowing them to quickly absorb and recall vast amounts of information, which can be advantageous in certain roles requiring data analysis or research.

By leveraging these strengths and providing necessary accommodations, organizations can foster an environment where neurodivergent individuals can thrive and contribute uniquely to the overall workforce innovation.

Promoting Diversity and Inclusion

To fully capitalize on the benefits of neurodiversity in workforce innovation, organizations must actively promote diversity and inclusion. This involves:

- **Creating Inclusive Hiring Practices:** Implementing inclusive recruitment and hiring practices is crucial for attracting and selecting neurodivergent talent. This includes modifying job descriptions, removing unnecessary prerequisites, and providing alternative pathways to employment, such as skills-based assessments or work trials.

- **Providing Reasonable Accommodations:** Accommodations play a vital role in ensuring equal opportunity and success for neurodivergent employees. These accommodations may include flexible work arrangements, sensory supports (e.g., noise-cancelling headphones), and clear communication channels.

- **Training and Sensitization Programs:** Educating the entire workforce about neurodiversity, its benefits, and best practices for creating an inclusive environment is crucial. This can be achieved through training programs, workshops, and informative materials that dispel myths and misconceptions about neurodivergent individuals.

- **Diversifying Leadership and Decision-Making Roles:** To truly embrace neurodiversity, organizations must ensure neurodivergent individuals have representation in leadership and decision-making positions. This helps foster an inclusive culture where diverse perspectives are valued.

- **Celebrating Neurodivergent Contributions:** Recognizing and celebrating the unique contributions of neurodivergent employees not only boosts their self-esteem but also helps create a culture of appreciation for diverse talents and perspectives.

By adopting these practices, organizations can tap into the full potential of their workforce, fostering an environment of collaboration and innovation.

Real-World Examples

Several companies have already embraced neurodiversity in their workforce and reaped the rewards of innovation. One such example is SAP, a multinational software corporation. SAP initiated the Autism at Work program, which specifically aims to recruit and retain neurodivergent talent. By providing individualized accommodations, job coaching, and opportunities for professional growth, SAP has witnessed significant improvements in productivity, creativity, and overall employee satisfaction.

Another notable example is Microsoft's Autism Hiring Program. Microsoft recognizes that neurodivergent individuals have unique capabilities and perspectives that add value to their teams. Their program focuses on inclusive hiring practices, on-the-job coaching, and ongoing support for neurodivergent employees. As a result, Microsoft has reported improved problem-solving, increased innovation, and enhanced team dynamics.

These success stories demonstrate the tangible benefits of embracing neurodiversity in the workforce and debunk the notion that neurodivergent individuals are a liability. Instead, organizations that prioritize diversity and inclusion enjoy a competitive advantage by fostering a culture of innovation and creativity.

Future Directions and Challenges

While there has been significant progress in promoting neurodiversity in the workplace, there are still challenges to overcome. Some of the future directions for neurodiversity in workforce innovation include:

- **Increasing Awareness and Education:** Continued efforts are needed to raise awareness about neurodiversity and its benefits. Educating employers, employees, and the general public about the unique strengths and capabilities of neurodivergent individuals can help dispel stereotypes and reduce stigma.

- **Expanding Access to Support Services:** Access to support services, such as job coaching, mentoring, and specialized training, is crucial for the successful integration and retention of neurodivergent employees. Organizations and policymakers should work together to ensure the availability and affordability of these resources.

- **Advocacy for Policy Changes:** Advocacy at both the organizational and governmental levels can drive policy changes that promote the inclusion of neurodivergent individuals in the workforce. This includes advocating for anti-discrimination laws, reasonable accommodations, and inclusive educational opportunities.

- **Collaborating with Research Institutions:** Collaborations between organizations and research institutions can further enhance our understanding of neurodiversity and its implications for workforce innovation. Research can inform evidence-based practices and interventions that support the success of neurodivergent employees.

By addressing these challenges and embracing neurodiversity in the workforce, organizations can create an inclusive and innovative environment that benefits both employees and the bottom line.

Conclusion

Embracing neurodiversity in workforce innovation brings immense benefits to organizations. The unique strengths and perspectives of neurodivergent individuals fuel creativity, problem-solving, and overall innovation. By promoting diversity and inclusion, providing necessary accommodations, and celebrating neurodivergent contributions, organizations can tap into the untapped potential of their workforce. Real-world examples from companies like SAP and Microsoft have demonstrated the positive impacts of embracing neurodiversity in the workplace. However, future directions and challenges, such as increasing awareness, expanding access to support services, advocating for policy changes, and collaborating with research institutions, must be addressed to fully unlock the value of diversity in workforce innovation. Ultimately, organizations that prioritize neurodiversity are not only fostering a culture of inclusion but also gaining a competitive advantage in today's rapidly changing business landscape.

Advocacy for Neurodivergent Employees

Advocacy plays a crucial role in promoting the rights and well-being of neurodivergent individuals in the workplace. It involves empowering and supporting employees with neurodivergent traits, challenging discriminatory practices, and promoting inclusive policies and practices. In this section, we will discuss the importance of advocacy for neurodivergent employees and explore strategies to effectively advocate for their needs and rights.

Understanding the Need for Advocacy

Neurodivergent individuals often face unique challenges in the workplace due to the way their brains are wired. They may have different communication styles, sensory sensitivities, or difficulty with social interactions. As a result, they may encounter barriers in finding and maintaining employment or experience discrimination and exclusion in the workplace.

Advocacy is essential in addressing these challenges and ensuring that neurodivergent employees have equal opportunities and support. It helps raise awareness about neurodiversity and challenges stereotypes and misconceptions. By advocating for inclusive policies, reasonable accommodations, and supportive work environments, we can foster a more equitable and inclusive workplace for all.

Strategies for Effective Advocacy

1. **Education and Awareness:** Advocacy begins with educating employers, colleagues, and other stakeholders about neurodiversity and the unique strengths and abilities of neurodivergent individuals. This can be done through workshops, presentations, or awareness campaigns. By increasing understanding and promoting a positive perception of neurodivergent employees, we can create a more inclusive work culture.

2. **Collaboration and Partnership:** Collaborating with organizations, community groups, and other stakeholders is crucial in advocating for the rights of neurodivergent employees. By joining forces, we can amplify our voices, share resources, and advocate for systemic changes. Partnering with disability advocacy groups, unions, or professional organizations can provide valuable support and guidance.

3. **Reasonable Accommodations:** One of the key aspects of advocacy is working to ensure that neurodivergent employees have access to reasonable accommodations that support their unique needs. This may include flexible work arrangements, modifications to the work environment, or assistive technologies. Advocates can engage with employers to raise awareness about the importance of accommodations and provide guidance on implementing them effectively.

4. **Policy Development:** Advocacy efforts should also focus on influencing policy development at the organizational or governmental level. This can involve advocating for inclusive hiring practices, anti-discrimination policies, and workplace diversity initiatives. By engaging in policy discussions and decision-making processes, advocates can contribute to creating a more inclusive and supportive environment for neurodivergent employees.

5. **Mentoring and Support Networks:** Establishing mentoring programs or support networks for neurodivergent employees can be an effective advocacy strategy. Providing opportunities for peer support, networking, and professional development can help build confidence and resilience among neurodivergent employees. Advocates can play a crucial role in facilitating the creation of these support systems within organizations.

6. **Promoting Neurodivergent Voices:** Advocacy should actively involve and amplify the voices of neurodivergent individuals themselves. By providing platforms for neurodivergent employees to share their experiences, insights, and ideas, advocates can empower them to contribute to the dialogue around neurodiversity in the workplace. This can be done through employee resource groups, speaking engagements, or participation in diversity and inclusion initiatives.

Case Study: Autism at Work Program

One successful example of advocacy for neurodivergent employees is the Autism at Work program launched by several major companies. This program aims to recruit and retain individuals with autism spectrum disorder (ASD) by providing tailored support and accommodations. Through partnerships with autism organizations, these companies have developed interview processes, training programs, and workplace adaptations to create a more inclusive environment. The program has demonstrated that with the right support, individuals with ASD can excel in diverse roles and contribute to the success of the organization.

Resources for Advocacy

Advocacy for neurodivergent employees is a collaborative and ongoing effort. To support individuals and organizations in their advocacy journey, here are some useful resources:

1. The National Autistic Society (NAS): The NAS provides guidance, best practices, and resources for employers, employees, and advocates promoting neurodiversity in the workplace.

2. Job Accommodation Network (JAN): JAN offers information, resources, and consulting services to help employers and employees navigate workplace accommodations and inclusivity.

3. Disability:IN: Disability:IN is a global nonprofit that promotes disability inclusion in business and drives strategies and policies for workplaces that are more inclusive of neurodivergent individuals.

4. Autistic Self Advocacy Network (ASAN): ASAN is a grassroots organization run by and for autistic individuals, advocating for inclusive policies and systems change that supports neurodivergent individuals.

5. Neurodiversity in the Workplace Toolkit: This toolkit, developed by the American Psychological Association, provides practical guidance and resources for employers and employees to create more inclusive workplaces.

Remember, advocacy is a continuous process, and it requires persistence, collaboration, and an ongoing commitment to promoting neurodiversity and inclusivity in the workplace. By advocating for neurodivergent employees, we can create a more equitable and accepting society where everyone can thrive.

Employment Challenges and Accommodations

Workplace Barriers and Discrimination

Understanding Workplace Barriers

In the context of neurodiversity, workplace barriers refer to the obstacles that neurodivergent individuals encounter in employment settings due to societal, organizational, and attitudinal factors. These barriers can hinder the full participation and inclusion of neurodivergent employees, leading to unequal access to opportunities, limited career advancement, and potential discrimination.

Workplace barriers can manifest in various forms, including:

1. Lack of Awareness and Understanding: One major barrier is the lack of awareness and understanding about neurodivergent conditions among employers and co-workers. This can result in misconceptions, stereotypes, and biases that contribute to a hostile or exclusionary work environment.

2. Inflexible Workplace Policies: Many workplaces have rigid policies and procedures that do not consider the unique needs of neurodivergent individuals. For example, strict attendance policies may not accommodate individuals who may need flexible work hours due to sensory sensitivities, therapy appointments, or other related reasons.

3. Limited Reasonable Accommodations: Reasonable accommodations are modifications or adjustments made to the work environment that enable individuals with disabilities, including neurodivergent individuals, to perform their job duties. However, there is often a lack of awareness and implementation of appropriate accommodations, such as providing noise-canceling headphones, flexible workspaces, or modified communication methods.

4. Social and Communication Challenges: Neurodivergent individuals may experience difficulties in social interactions and communication, which can lead to misunderstandings, isolation, and exclusion in the workplace. Challenges in non-verbal communication, understanding social cues, and initiating or maintaining conversations may be perceived as a lack of competence or professionalism.

5. Sensory Overload and Environmental Factors: Many neurodivergent individuals have sensory sensitivities, meaning they may be more sensitive to certain stimuli, such as bright lights, loud noises, or strong smells. Workplaces that do not consider these sensory needs can contribute to heightened stress, anxiety, and reduced productivity for neurodivergent employees.

6. Implicit Bias and Discrimination: Implicit biases are unconscious stereotypes or prejudices that influence our judgments and behaviors. Neurodivergent individuals may face discrimination in the workplace due to biases against their condition or misunderstandings about their abilities. They may be overlooked for promotions, underestimated in their job performance, or face negative stereotypes that perpetuate exclusionary practices.

Addressing Workplace Barriers

Addressing workplace barriers and discrimination is crucial for creating an inclusive and supportive environment for neurodivergent employees. Here are some strategies that can help promote neurodiversity acceptance and inclusion in the workplace:

1. Education and Awareness: Employers must invest in education and awareness programs to educate staff about neurodiversity, including specific neurodivergent conditions and the strengths and challenges associated with them. This can help mitigate misconceptions, reduce biases, and foster a culture of understanding and acceptance.

2. Reasonable Accommodations: Employers should engage in a collaborative process with neurodivergent employees to identify and implement reasonable accommodations tailored to their individual needs. Accommodations can include flexible work schedules, quiet workspaces, clear communication guidelines, and assistive technologies. Providing these accommodations not only benefits the neurodivergent employees but also enhances the overall productivity and success of the workplace.

3. Inclusive Policies and Practices: Employers should review and revise existing policies and practices to ensure they are inclusive and considerate of neurodivergent needs. This can include flexibility in work hours, communication styles, and performance evaluation criteria. Additionally, incorporating diversity and

inclusion initiatives and ensuring equitable access to professional development opportunities can help create a more inclusive and supportive work environment.

4. Sensory-Friendly Workspaces: Employers can create sensory-friendly workspaces by making modifications to the physical environment. This may involve providing adjustable lighting, quiet areas, or implementing a sensory-friendly design in common areas. Taking steps to minimize sensory distractions can reduce stress and enhance the comfort and productivity of neurodivergent employees.

5. Training and Sensitivity Workshops: Employers should provide training and sensitivity workshops to all employees to develop an understanding of neurodiversity and foster empathy and respect towards neurodivergent individuals. These workshops can cover topics such as effective communication strategies, recognition of talents and strengths, and dispelling misconceptions.

6. Supportive Work Culture: A supportive work culture is essential for the well-being and success of neurodivergent employees. Employers should promote a culture of respect, open communication, and inclusivity. Encouraging diversity and fostering an environment where all employees are valued and can excel contributes to the overall success of the organization.

By addressing these workplace barriers and fostering a culture of acceptance, organizations can tap into the unique strengths and talents that neurodivergent individuals bring to the table.

Real-World Example

A concrete example of a workplace addressing these barriers is the software company SAP (Systems, Applications, and Products). SAP recognizes the value of neurodivergent talent and has created the Autism at Work program, which actively recruits and employs individuals on the autism spectrum. The program offers a supportive work environment with tailored onboarding, mentorship, and accommodations.

In collaboration with autism advocacy organizations, SAP provides training to their employees to raise awareness about autism and promote inclusive practices. Through this program, the company has found that neurodivergent individuals can excel in tasks that require attention to detail, pattern recognition, and logical thinking.

SAP's Autism at Work program demonstrates how organizations can successfully address workplace barriers and harness the unique abilities of neurodivergent individuals to drive innovation and business success.

Further Resources

For further reading on workplace barriers and discrimination, the following resources can provide valuable insights and information:

1. Book: "Neurodiversity in the Workplace: Strength-Based Strategies to Help Employees with Autism Spectrum Disorder Succeed" by Marilyn Bisberg and Paula Moraine.

2. Research Article: "Neurodiverse Talent and Cultural Diversity in the Workplace: A Review" by Elektra Papadopoulos, et al. (Journal of Occupational Psychology, 2020).

3. Website: The National Autistic Society (www.autism.org.uk) provides resources and guidance for supporting neurodivergent individuals in the workplace.

4. Webinar: The TED Talk "The Power of Neurodiversity" by Dr. Robert D. Austin highlights the value of neurodivergent perspectives in the business world.

Remember that creating an inclusive work environment requires ongoing commitment and continuous improvement. By adopting these strategies, organizations can pave the way for a more inclusive and diverse workforce, benefiting both neurodivergent individuals and the broader community.

Creating Inclusive Hiring Practices

In order to foster a diverse and inclusive workplace, it is essential for organizations to establish inclusive hiring practices. This section will discuss the key principles and strategies for creating an inclusive hiring process that welcomes and supports neurodivergent individuals.

Understanding Neurodivergent Talent

Neurodivergent individuals bring unique strengths and perspectives to the workplace. It is important for employers to understand the range of neurodivergent conditions and the corresponding talents they may possess. Some common neurodivergent conditions include Autism Spectrum Disorder (ASD), Attention-Deficit/Hyperactivity Disorder (ADHD), dyslexia, and intellectual disabilities. Each condition comes with its own set of strengths and challenges, and recognizing and valuing these strengths is crucial for building an inclusive workforce.

For instance, individuals with ASD often exhibit exceptional attention to detail, strong analytical skills, and creativity. Those with ADHD may possess high energy levels, the ability to multitask, and a fresh, innovative perspective. Dyslexic individuals may have well-developed problem-solving skills, strong visual-thinking abilities, and a keen eye for detail. Intellectual disabilities should not be seen as barriers, as individuals with these conditions often demonstrate exceptional interpersonal skills, reliability, and dedication.

By understanding and appreciating the unique strengths of neurodivergent individuals, organizations can tap into a pool of talent that brings fresh ideas, different approaches to problem-solving, and enhanced diversity of thought.

Reevaluating Job Descriptions and Requirements

Traditional job descriptions and requirements can inadvertently create barriers for neurodivergent individuals. To attract a diverse pool of candidates, it is important to revise and reframe job descriptions using inclusive language and criteria. Instead of focusing solely on specific qualifications and experiences, consider highlighting desirable skills and competencies that can be acquired through various paths.

For example, rather than requesting a specific number of years of experience in a certain field, emphasize transferable skills such as critical thinking, problem-solving, adaptability, and teamwork. By broadening the criteria and focusing on essential competencies, employers can ensure that neurodivergent candidates with non-traditional backgrounds are not overlooked.

Diverse Sourcing and Recruitment Strategies

Creating inclusive hiring practices involves expanding recruitment efforts to reach a more diverse pool of candidates. Traditional recruitment channels may not effectively reach neurodivergent individuals, so it is important to actively seek out alternative sourcing methods.

Establishing partnerships with neurodivergent organizations, disability-focused employment agencies, and community-based programs can help connect employers with neurodivergent talent. These partnerships can facilitate outreach, provide guidance on inclusive hiring practices, and offer support throughout the recruitment process.

Additionally, leveraging online platforms, social media, and job boards that specifically cater to neurodivergent individuals can help attract a broader range of candidates. This includes platforms like Neurodiversity Connect, Autistic Employment Network, and other disability-focused job sites.

Accessible Application and Interview Processes

It is crucial to ensure that the application and interview processes are accessible to neurodivergent candidates. Companies should strive to eliminate any unnecessary barriers and create an inclusive experience for all applicants. Some strategies to consider include:

- Providing clear instructions and guidelines: Clearly outline the application process and provide instructions in a straightforward, concise manner. Avoid using jargon or overly complex language that may confuse or exclude neurodivergent individuals.

- Offering flexibility in application methods: Provide multiple options for candidates to submit their applications, such as email, online forms, or video submissions. This allows neurodivergent individuals to choose the method that suits their preferred communication style and accommodates their specific needs.

- Preparing candidates for interviews: Inclusive hiring practices involve making accommodations for neurodivergent candidates during interviews. Inform candidates about the format and structure of the interview in advance, and offer additional time or alternative formats if needed. Consider providing interview questions in advance to allow neurodivergent individuals to prepare and showcase their abilities effectively.

- Evaluating candidates holistically: In the selection process, it is important to take a holistic view of candidates' skills and abilities. Focus on the alignment between the candidate's strengths and the job requirements, rather than solely relying on interview performance or conventional markers of success.

Training and Sensitization

To ensure the successful integration of neurodivergent employees into the workplace, it is essential to provide training and sensitization to hiring managers, interviewers, and the broader workforce. This can help create an inclusive and supportive environment for all employees.

Training should cover topics such as understanding neurodivergent conditions, recognizing and appreciating neurodivergent strengths, effective communication strategies, and behavioral adjustments that may be needed to accommodate diverse individuals. It is important to promote a culture of understanding, empathy, and respect throughout the organization.

Establishing Supportive Workplace Accommodations

To enable the success of neurodivergent employees, organizations should provide supportive workplace accommodations. These accommodations can include:

- Flexible work schedules: Offering flexible work hours or remote work options can help neurodivergent individuals manage their unique needs and enhance their productivity.

- Sensory-friendly environments: Creating a workspace that considers sensory sensitivities can significantly improve the well-being and comfort of neurodivergent employees. This can involve providing noise-cancelling headphones, offering designated quiet areas, or adjusting lighting levels.

- Mentoring and coaching: Pairing neurodivergent employees with mentors or coaches can provide guidance, support, and help navigate any challenges they may face in the workplace.

- Communication and social support: Encouraging clear and direct communication, as well as fostering a culture of inclusivity and respect, can help neurodivergent individuals feel valued and supported. Offering social skills training or peer support programs can also facilitate social integration.

- Job carving and task customization: Tailoring job roles and responsibilities to align with the strengths and interests of neurodivergent employees can maximize their potential and job satisfaction. By allowing for task customization, companies can leverage the unique skills that neurodivergent individuals bring to the table.

Measurement and Evaluation

To ensure the effectiveness of inclusive hiring practices, organizations should establish measurement and evaluation systems. This involves tracking and analyzing relevant data to identify areas for improvement and ensure ongoing commitment to diversity and inclusion.

Metrics can include the number of neurodivergent employees hired, retention rates, employee satisfaction surveys, feedback from neurodivergent employees, and the impact on workplace culture. Regular evaluation and iteration of inclusive hiring practices will help organizations refine their approach and make informed decisions.

Conclusion

Creating inclusive hiring practices is not only a moral imperative but also a strategic advantage for organizations seeking to build diverse and innovative workplaces. By understanding and valuing the unique strengths that neurodivergent individuals bring, organizations can take proactive steps to attract, support, and retain neurodiverse talent. Embracing diversity and neurodiversity in the hiring process fosters a culture of innovation, enhances problem-solving capabilities, and paves the way for a more inclusive future.

Accessibility and Accommodations for Neurodivergent Individuals

In order to create an inclusive environment for neurodivergent individuals, it is crucial to provide accessibility and accommodations that cater to their unique needs. These accommodations can greatly enhance their participation, engagement, and overall well-being in various settings, including education and the workplace. In this section, we will explore the importance of accessibility and accommodations, as well as discuss some specific strategies that can be implemented to support neurodivergent individuals.

Understanding Neurodivergent Needs

Neurodivergent individuals often experience difficulties in processing sensory information, social interactions, communication, and executive functioning. These challenges can vary among different neurodivergent conditions, such as autism, ADHD, dyslexia, and intellectual disabilities. Therefore, it is essential to have a comprehensive understanding of the specific needs of each individual.

Physical Environment Modifications

One key aspect of accessibility for neurodivergent individuals is modifying the physical environment to minimize sensory overload and provide a comfortable space for them to thrive. This can include:

- Creating quiet areas or sensory-friendly spaces where individuals can retreat when overwhelmed.
- Providing noise-cancelling headphones or earplugs to reduce auditory distractions.

- Using adjustable lighting options to accommodate individuals' sensitivities to bright or flickering lights.
- Designing visually clean and clutter-free spaces to reduce visual stimuli.
- Implementing sensory-friendly seating arrangements, such as flexible seating options or designated areas for movement.

By making these modifications, the physical environment becomes more inclusive and conducive to neurodivergent individuals' needs.

Assistive Technologies

Assistive technologies can play a vital role in supporting neurodivergent individuals in their daily activities, education, and work. These technologies can help compensate for challenges in communication, organization, and information processing. Examples of assistive technologies include:

- Augmentative and alternative communication (AAC) devices, such as speech-generating devices and communication apps, to support individuals with limited verbal skills.
- Text-to-speech and speech-to-text software to facilitate reading, writing, and communication for individuals with dyslexia or language difficulties.
- Visual organizers and timers to assist individuals with ADHD or executive functioning challenges in planning and time management.
- Digital tools and apps that provide personalized learning experiences and accommodate different learning styles and preferences.

These assistive technologies can empower neurodivergent individuals to overcome their specific challenges and enhance their independence and engagement.

Flexible Instructional Strategies

In educational settings, using flexible instructional strategies can significantly benefit neurodivergent learners. These strategies allow for personalized learning experiences that cater to individual strengths and challenges. Some examples include:

- Differentiating instruction to accommodate diverse learning needs and preferences.
- Providing visual supports, such as visual schedules, graphic organizers, and visual cues, to enhance understanding and organization for individuals with autism or ADHD.
- Breaking down complex tasks into smaller, manageable steps to support individuals with executive functioning difficulties.
- Incorporating multisensory learning activities to engage different modes of sensory input and enhance learning retention for individuals with dyslexia or sensory processing difficulties.
- Allowing flexible seating options, movement breaks, and other sensory supports to promote self-regulation and attention for individuals with ADHD or sensory sensitivities.

By adopting these instructional strategies, educators can create inclusive learning environments that meet the diverse needs of neurodivergent students.

Accommodation Plans and Individualized Supports

To ensure that neurodivergent individuals receive the necessary accommodations, it is important to develop individualized accommodation plans in collaboration with the individual, their families, and relevant professionals. These plans should be tailored to meet specific needs and outline the necessary supports and accommodations in various contexts.

Examples of accommodations and supports that can be included in these plans are:

- Extended time for assignments, tests, or exams to account for difficulties in processing information or completing tasks.
- Modified assignments or alternative assessment formats that align with individuals' strengths and learning styles.
- Preferential seating to minimize distractions and promote engagement.
- Access to a quiet space or a designated support person in high-stress situations.

- Regular check-ins with a designated mentor or counselor to address emotional and social well-being.

By implementing individualized accommodation plans, individuals can access the necessary supports to succeed academically and holistically.

Advocacy and Collaboration

Promoting accessibility and accommodations for neurodivergent individuals requires collaboration and advocacy at various levels. This includes:

- Educating and raising awareness among educators, employers, colleagues, and the general public about the importance of accessibility and accommodations for neurodivergent individuals.
- Collaborating with individuals, families, educators, professionals, and advocacy organizations to develop inclusive practices and policies.
- Providing training and professional development opportunities for educators, employers, and support staff to enhance their understanding of neurodiversity and effective strategies for accommodations.
- Involving neurodivergent individuals in decision-making processes to ensure their voices and perspectives are heard and respected.

By actively advocating for accessibility and accommodations, we can create more inclusive environments that celebrate neurodiversity.

Resources and Support Networks

Neurodivergent individuals and their families can benefit from access to resources and support networks that offer guidance, information, and community connections. These resources can include:

- Websites, books, and online communities that provide information and practical tips on accommodations and accessibility.
- Support groups and networks where individuals can connect with peers and share experiences and strategies.
- Professional organizations and advocacy groups that offer guidance, resources, and workshops on navigating accommodations and promoting neurodiversity.

- Accessible mental health services that address the specific emotional and psychological needs of neurodivergent individuals.

By leveraging these resources and support networks, individuals can find the necessary guidance and connections to navigate their unique journeys.

Conclusion

Accessibility and accommodations are essential for creating an inclusive and supportive environment for neurodivergent individuals. By modifying physical environments, implementing assistive technologies, adopting flexible instructional strategies, developing individualized accommodation plans, advocating for change, and providing necessary resources and support networks, we can promote the inclusion and well-being of neurodivergent individuals in education, the workplace, and society at large. Through these efforts, we celebrate neurodiversity and empower each person to thrive in their own unique way.

% References

Bibliography

[1] Anderson, J. (2017). Supporting students with autism spectrum disorder in higher education: Current practices and trends. *Journal of Autism and Developmental Disorders, 47*(8), 2519-2533.

[2] Armstrong, J. (2019). *Neurodiversity in the classroom: Strength-based strategies to help students with special needs succeed in school and life.* Alexandria, VA: ASCD.

[3] Grandin, T., & Panek, R. (2013). *The autistic brain: Thinking across the spectrum.* Boston, MA: Mariner Books.

[4] Higashida, N., & Yoshida, K. (2019). *Fall down seven times, get up eight: A young man's voice from the silence of autism.* New York, NY: Random House Trade Paperbacks.

[5] Myles, B. S., Adreon, D., & Gitlitz, D. (2001). *Autism spectrum disorders in adolescence and adulthood: Research, diagnosis, and practice.* Austin, TX: PRO-ED.

Neurodiversity-Friendly Job Training Programs

Neurodiversity-friendly job training programs play a crucial role in supporting the inclusion and success of neurodivergent individuals in the workplace. These programs aim to address the specific strengths and challenges faced by neurodivergent individuals, providing them with the necessary skills and support to thrive in their chosen career paths. In this section, we will explore the key principles and components of neurodiversity-friendly job training programs, as well as highlight some successful examples and strategies.

Understanding the Needs of Neurodivergent Individuals

Effective job training programs for neurodivergent individuals start by understanding their unique needs and considering the diversity within the population. It is crucial to recognize that neurodivergent individuals have a wide range of strengths, abilities, and preferences. Therefore, a one-size-fits-all approach is not suitable.

To cater to the diverse needs of neurodivergent individuals, job training programs should adopt a person-centered approach. This involves taking into account the individual's interests, strengths, and learning styles. By understanding and leveraging their unique strengths, such as exceptional attention to detail, pattern recognition, or problem-solving abilities, job training programs can empower neurodivergent individuals to excel in their chosen field.

Collaborative and Multidisciplinary Approach

Neurodiversity-friendly job training programs recognize the importance of collaboration and a multidisciplinary approach. They involve a team of professionals, including job coaches, occupational therapists, psychologists, and educators, who work together to support neurodivergent individuals in acquiring and applying job-related skills.

The multidisciplinary team collaboratively develops and implements individualized training plans to meet the specific needs of each neurodivergent individual. This may include explicit instruction, modeling, hands-on practice, and consistent feedback. Additionally, experts in assistive technology can provide guidance on the effective use of tools and technologies that can enhance job performance and productivity.

Building a Supportive Environment

Creating a supportive and inclusive environment is key to the success of neurodiversity-friendly job training programs. This involves fostering a culture of acceptance, understanding, and empathy among employers, coworkers, and other stakeholders.

Employers and colleagues should receive training and awareness programs to increase their understanding of neurodivergent individuals and their needs. By promoting empathy and eliminating biases and stereotypes, employers can create an environment where neurodivergent individuals feel valued and supported.

To ensure a smooth transition into the workplace, job training programs should also offer ongoing support, including mentorship programs and peer support networks. These initiatives can provide guidance, advice, and social connections that help neurodivergent individuals navigate the challenges they may face in the workplace.

Tailoring Training Methods

Neurodiversity-friendly job training programs should adopt training methods that align with the cognitive and learning styles of neurodivergent individuals. Traditional lecture-based approaches may not be effective for everyone, especially those who have difficulties with attention, sensory processing, or executive functioning.

Alternative training methods, such as visual supports, step-by-step instructions, and hands-on learning opportunities, can enhance comprehension and skill acquisition for neurodivergent individuals. Additionally, incorporating technology, such as virtual reality or interactive simulations, can provide valuable experiential learning opportunities that bridge the gap between theory and practice.

Transition Support

Smooth transition into employment is critical for the success of neurodivergent individuals. Job training programs should not only focus on skill development but should also provide support during the transition phase. This may include assistance with job applications, interview preparation, disclosure of neurodivergent identity, and workplace socialization.

Moreover, collaboration between job training programs and potential employers can facilitate the creation of supportive work environments. Internship programs or job-shadowing opportunities can provide practical experience and help neurodivergent individuals explore different career options while building their professional networks.

Success Stories and Best Practices

Numerous neurodiversity-friendly job training programs have been successful in empowering neurodivergent individuals and promoting their inclusion in the workforce.

For example, the Autism Advantage program, developed by the Autistic Self Advocacy Network (ASAN) and JPMorgan Chase, provides comprehensive training and support for individuals on the autism spectrum. The program focuses on developing key job skills, including communication, teamwork, and problem-solving, while also addressing the social and sensory needs of participants.

Another notable program is the Neurodiversity at Work initiative by Microsoft. This program aims to recruit and support neurodivergent individuals in various job roles. It offers a structured onboarding process, ongoing job coaching, and a supportive community that fosters inclusion and growth.

These success stories highlight the importance of tailoring job training programs to the needs of neurodivergent individuals and creating supportive environments that value their unique strengths and contributions.

Resources and Tools

There are several resources and tools available to support the development and implementation of neurodiversity-friendly job training programs:

- Job Accommodation Network (JAN): JAN provides free consulting services and resources to help employers and job seekers understand and implement workplace accommodations that support individuals with disabilities, including neurodivergent individuals.

- National Collaborative on Workforce and Disability for Youth (NCWD/Youth): NCWD/Youth offers a range of resources and publications focused on promoting successful employment outcomes for youth with disabilities, including toolkits and guides for job training programs.

- Assistive Technology: Tools such as text-to-speech software, task management apps, and noise-canceling headphones can enhance accessibility and support the productivity of neurodivergent individuals in the workplace.

- Professional Associations and Advocacy Organizations: Organizations like the Autistic Self Advocacy Network (ASAN), Autism Society, and Attention Deficit Disorder Association (ADDA) provide valuable resources, advocacy support, and networking opportunities for job training programs and employers.

Conclusion

Neurodiversity-friendly job training programs play a vital role in promoting the inclusion and success of neurodivergent individuals in the workplace. By understanding their unique needs, adopting a person-centered approach, and creating supportive environments, these programs can empower neurodivergent individuals to thrive in their chosen careers. Collaboration, tailored training methods, and transition support are key components of effective job training programs. Success stories and resources provide valuable guidance and inspiration for creating and implementing neurodiversity-friendly job training programs. With continued efforts and awareness, we can build a more inclusive workforce that celebrates the talents and contributions of all individuals, neurodivergent or otherwise.

Addressing Sensory Overload and Workplace Environment

Sensory overload is a common experience for many individuals with neurodivergent conditions, such as autism spectrum disorder (ASD) and attention-deficit/hyperactivity disorder (ADHD). It occurs when there is an overwhelming amount of sensory information for the brain to process, leading to feelings of stress, anxiety, and discomfort. In the workplace, sensory overload can significantly impact the well-being and productivity of neurodivergent employees. In this section, we will explore strategies to address sensory overload and create a sensory-friendly workplace environment.

Understanding Sensory Overload

Sensory overload can occur across various sensory domains, including auditory, visual, olfactory, tactile, and proprioceptive. Neurodivergent individuals may have different sensory processing profiles, with some being hypersensitive (sensitive to sensory stimuli) and others being hyposensitive (less sensitive to sensory stimuli). Understanding the individual's sensory needs and preferences is crucial in creating a supportive workplace environment.

Environmental Modifications

Creating a sensory-friendly workplace environment involves making modifications to the physical space to reduce sensory stimuli that may trigger overload. Here are some strategies to consider:

- **Noise reduction:** Minimize background noise by using sound-absorbing materials, providing noise-canceling headphones, or creating designated quiet areas for employees to retreat to when needed.

- **Lighting adjustments:** Use adjustable lighting options to accommodate individual preferences. Some individuals may prefer dimmer lighting, while others may benefit from full-spectrum lighting. Natural lighting, when possible, can also contribute to a more comfortable environment.

- **Color choices:** Opt for neutral or calming colors on walls and furniture to create a soothing ambiance. Avoid bright and visually stimulating colors that may be overwhelming for some individuals.

- **Temperature control:** Maintain a comfortable and consistent temperature in the workspace. Extremes in hot or cold environments can be distracting and impact focus.

Flexible Workstations

Providing flexible workstations can allow employees to tailor their immediate work environment to their sensory needs. Consider the following adaptations:

- **Ergonomic furniture:** Provide adjustable chairs, desks, and monitor stands to accommodate different body postures and preferences. This ensures physical comfort and reduces physical distractions.

- **Personalization options:** Allow employees to bring in items that provide sensory comfort, such as stress balls, fidget toys, or personal noise-cancelling devices. Giving individuals control over their immediate environment can empower them to manage sensory inputs effectively.

- **Flexible seating arrangements:** Offer different seating options, such as standing desks or alternative seating (e.g., stability balls or bean bags), to cater to individual preferences and promote movement.

Sensory Break Spaces

Sensory break spaces are designated areas where employees can go to relax, recharge, and manage sensory overload. These spaces should be designed to provide a calm and sensory-friendly environment. Consider the following elements when creating sensory break spaces:

- **Comfortable seating:** Provide soft seating options, such as bean bags or padded chairs, for individuals to relax and unwind.

- **Sensory regulation tools:** Include sensory tools like weighted blankets, noise-canceling headphones, stress balls, or sensory swings to support self-regulation and provide comfort.
- **Low-stimulation environment:** Ensure the area has minimal visual distractions, with soft lighting and muted colors. Implement measures to reduce auditory stimuli, such as soundproofing or a white noise machine.
- **Privacy considerations:** Create spaces that offer privacy and allow individuals the option to retreat from social interactions when needed.

Sensory Awareness and Training

Educating both neurodivergent and neurotypical employees about sensory differences and sensory overload can foster a more inclusive and empathetic work environment. Consider the following strategies:

- **Sensory awareness training:** Provide training sessions that increase awareness and understanding of sensory processing differences. This can help employees recognize and accommodate the needs of their neurodivergent colleagues.
- **Communication tools:** Develop tools, such as visual cues or sensory preference questionnaires, to facilitate open communication between employees. This allows individuals to express their sensory needs and preferences effectively.
- **Support networks:** Establish support networks or employee resource groups that bring together neurodivergent individuals to share experiences, insights, and strategies for managing sensory overload in the workplace.

Case Study: Sensory-Friendly Office Design

To illustrate the practical application of addressing sensory overload and creating a sensory-friendly workplace, let's consider the case of a software development company.

The company decides to renovate its office space, taking into account the unique sensory needs of its neurodivergent employees. They implement the following changes:

- **Soundproofing:** The company installs acoustic panels and carpets to absorb excess noise and create a quieter work environment.

- **Adjustable lighting:** Each workstation is equipped with adjustable desk lamps, allowing employees to customize their lighting preferences. Natural lighting is maximized by providing large windows.
- **Flexible workstations:** Ergonomic chairs and height-adjustable desks are provided to promote physical comfort and reduce postural strain.
- **Sensory break spaces:** The company creates several sensory break rooms, each featuring comfortable seating, calming decor, and sensory regulation tools like fidget toys and noise-cancelling headphones.
- **Sensory awareness training:** The HR department conducts sensory awareness training for all employees to foster a better understanding of sensory differences and promote a supportive work culture.

The sensory-friendly changes result in a more inclusive and accommodating work environment, where neurodivergent employees feel supported and empowered to thrive.

In conclusion, addressing sensory overload and creating a sensory-friendly workplace is essential for promoting the well-being and productivity of neurodivergent employees. By understanding sensory needs, making environmental modifications, providing flexible workstations, and offering sensory break spaces, organizations can create an inclusive work environment that celebrates neurodiversity and benefits all employees.

Workplace Supports and Policies

Employee Resource Groups (ERGs)

Employee Resource Groups (ERGs) play a crucial role in promoting inclusivity and supporting neurodivergent individuals in the workplace. ERGs are voluntary, employee-led organizations that bring together individuals with shared characteristics or experiences. In the context of neurodiversity, ERGs provide a space for employees who are neurodivergent, or who have family members or friends who are neurodivergent, to connect, share experiences, and advocate for their needs and rights within the organization.

ERGs function as a platform for raising awareness and fostering a culture of acceptance and support. They provide opportunities for networking, mentoring, and professional development, empowering neurodivergent employees to thrive and reach their full potential. In this section, we will explore the role of ERGs in promoting neurodiversity in the workplace, the benefits they offer, and how organizations can effectively establish and support these groups.

Benefits of ERGs for Neurodivergent Employees

ERGs provide several benefits to neurodivergent employees, enhancing their sense of belonging, well-being, and professional growth.

Firstly, ERGs create a supportive community where individuals can share their experiences and challenges unique to neurodivergent traits. This sense of belonging and understanding helps reduce feelings of isolation and promotes psychological well-being. Employees can find comfort and support in discussing their successes, obstacles, and strategies for navigating the workplace.

Secondly, ERGs foster opportunities for mentoring and career development. Neurodivergent employees can benefit from the guidance and advice of more experienced colleagues who have navigated similar situations. Mentorship programs within ERGs can help enhance professional skills, provide insights for career advancement, and build valuable relationships.

Thirdly, ERGs play a vital role in advocating for inclusive policies and accommodations within the organization. By collaborating with management and HR departments, ERGs can ensure that the needs of neurodivergent employees are understood and met. This includes advocating for workplace adjustments, such as flexible work hours, modified communication styles, and sensory-friendly environments, that can enhance productivity and well-being.

Establishing and Supporting ERGs

Establishing and supporting ERGs requires a collaborative approach between employees, management, and HR departments. Here are some key considerations for effectively setting up and sustaining these groups:

1. **Leadership Support:** Leaders within the organization should endorse and actively support the formation of ERGs. Their buy-in is crucial in providing the necessary resources, budget, and visibility to ensure the success of these groups.

2. **Employee Involvement:** ERGs should be open to all employees, whether they are neurodivergent themselves or allies. Encouraging diverse participation fosters a broader understanding of neurodiversity and creates a sense of unity within the organization.

3. **Clear Mission and Goals:** ERGs should establish a clear mission statement and define their goals and objectives. This clarity helps guide the activities and initiatives undertaken by the group, ensuring they align with the needs and priorities of neurodivergent employees.

4. **Collaboration and Partnerships:** ERGs can collaborate with other diversity and inclusion initiatives within the organization to leverage resources and amplify their impact. Building partnerships with HR departments, employee assistance programs (EAPs), and disability resource centers (DRCs) can provide access to expertise and support.

5. **Training and Education:** ERGs should prioritize training and education initiatives to raise awareness about neurodiversity among employees. This can include workshops, webinars, and guest speaker sessions that address common misconceptions, promote understanding, and encourage empathy.

6. **Resource Allocation:** Organizations should allocate sufficient resources to support ERGs, including funding for events, training, and professional development opportunities. Appointing a designated point person within HR or diversity departments to provide guidance and support can also help ensure the sustainability of the groups.

Case Study: ERG Success Story

To illustrate the impact of ERGs, let's consider the case study of a technology company that established an ERG specifically for neurodivergent employees. The ERG, named "NeuroCONNECT," aimed to create a supportive network, advocate for inclusive policies, and promote the hiring and retention of neurodivergent talent.

Under the NeuroCONNECT umbrella, the ERG organized various initiatives. One of their successful programs was a mentorship program that paired neurodivergent employees with senior leaders from different departments. The mentors provided guidance, coaching, and professional development opportunities to help neurodivergent employees navigate their careers within the organization.

Additionally, NeuroCONNECT collaborated with the HR department to implement a neurodiversity-friendly recruitment and interview process. They advocated for greater consideration of alternative employment assessments, such as skills-based evaluations, to ensure fair and unbiased evaluations of candidates.

The ERG also organized awareness campaigns during Autism Acceptance Month, including lunch-and-learn sessions, panel discussions, and art exhibitions featuring neurodivergent employees. These initiatives helped educate the broader employee population about neurodiversity and fostered a more inclusive work environment.

Through their efforts, NeuroCONNECT improved the workplace experience for neurodivergent employees and helped pave the way for organizational changes that benefit all employees. Their success underscores the power of ERGs in driving positive change, fostering inclusivity, and maximizing the potential of neurodivergent talent in the workplace.

In conclusion, ERGs play a vital role in promoting neurodiversity in the workplace. By providing a sense of community, supporting professional development, and advocating for inclusive policies, these groups empower neurodivergent employees to thrive. Organizations can establish and support ERGs by securing leadership buy-in, fostering diverse participation, defining clear mission statements, building collaborations, prioritizing training, and allocating resources. The impact of ERGs is exemplified by success stories like NeuroCONNECT, demonstrating the tangible benefits of these groups in creating inclusive, supportive, and thriving workplaces.

Mentoring and Coaching Programs for Neurodivergent Employees

Neurodivergent individuals bring unique perspectives, skills, and talents to the workplace. However, they may face certain challenges related to communication, social interactions, and sensory sensitivities. To support their success and promote inclusion, organizations have developed mentoring and coaching programs specifically tailored to the needs of neurodivergent employees.

Understanding Mentoring and Coaching

Mentoring and coaching programs provide guidance, support, and professional development opportunities to individuals in the workplace. While both serve similar purposes, there are slight differences between the two:

- **Mentoring:** Mentoring involves a partnership between an experienced employee (mentor) and a less experienced employee (mentee) who can benefit from their knowledge and guidance. Mentors act as role models, offering advice, sharing experiences, and helping mentees develop their skills and achieve their goals.
- **Coaching:** Coaching focuses on supporting the personal and professional growth of an individual. Coaches help employees identify and overcome challenges, enhance their performance, and develop specific skills. Coaching typically occurs in short-term, goal-oriented sessions, where the coach provides feedback and facilitates the mentee's self-discovery and learning.

For neurodivergent employees, mentoring and coaching programs can play a crucial role in facilitating their integration into the workplace and unlocking their full potential.

Benefits of Mentoring and Coaching for Neurodivergent Employees

Mentoring and coaching programs offer numerous benefits to neurodivergent employees, including:

- **Enhanced Social Skills:** Many neurodivergent individuals may struggle with social interactions, making it challenging to establish relationships and navigate workplace dynamics. Mentoring and coaching programs provide a supportive environment to practice and improve social skills, such as active listening, effective communication, and conflict resolution.
- **Improved Job Performance:** Neurodivergent employees often have unique strengths and abilities, such as attention to detail, problem-solving skills, and analytical thinking. Mentors and coaches can help them leverage these strengths to excel in their roles, enhancing their job performance and contributing to the organization's success.
- **Increased Confidence:** Neurodivergent employees may face self-doubt and anxiety in the workplace, particularly in unfamiliar or high-pressure

situations. Mentors and coaches provide encouragement, constructive feedback, and emotional support, helping neurodivergent employees build confidence, overcome challenges, and navigate their career paths.

Designing Effective Mentoring and Coaching Programs

To create effective mentoring and coaching programs for neurodivergent employees, organizations should consider the following strategies:

- **Individualized Approach:** Recognize that neurodivergent employees have unique needs, strengths, and learning styles. Tailor mentoring and coaching programs to accommodate their specific requirements and provide personalized support.
- **Matching Mentors and Mentees:** Pair neurodivergent employees with mentors who have experience or knowledge relevant to their specific challenges and goals. Consider factors such as personality, interests, and career aspirations to ensure a successful and productive mentoring relationship.
- **Training for Mentors and Coaches:** Provide mentors and coaches with training and resources on neurodiversity, the challenges faced by neurodivergent individuals, and strategies for effective communication and support. This training will enhance their ability to understand and meet the mentees' needs.
- **Regular Check-Ins and Feedback:** Establish regular check-in meetings between mentors or coaches and neurodivergent employees. These sessions provide opportunities to address concerns, evaluate progress, and offer guidance and feedback. Consistent communication is essential for building trust and ensuring mentees' ongoing support.
- **Peer Support Networks:** Alongside mentoring and coaching programs, create peer support networks for neurodivergent employees. These networks foster a sense of community, enable knowledge sharing, and provide a safe and inclusive space for discussing challenges and seeking advice.
- **Evaluation and Continuous Improvement:** Regularly evaluate the effectiveness of mentoring and coaching programs for neurodivergent employees. Collect feedback from both mentors and mentees to identify areas for improvement and make necessary adjustments to ensure the programs meet the evolving needs of neurodivergent employees.

Case Study: Autism Inclusion Program

One example of a successful mentoring and coaching program for neurodivergent employees is the Autism Inclusion Program implemented by Company X. The program aims to create a supportive and inclusive work environment for employees with Autism Spectrum Disorder (ASD) by providing them with mentors and coaches who have experience in working with neurodivergent individuals.

The mentors in the program undergo specialized training on understanding ASD, communication strategies, and best practices for supporting neurodivergent employees. They are matched with mentees based on their shared interests and career goals. Mentors meet regularly with their mentees to discuss work-related challenges, provide guidance, and help them develop their skills.

In addition to the mentoring component, the program also offers coaching sessions focused on enhancing social skills, self-advocacy, and personal development. Coaches work closely with the mentees to set goals, track progress, and provide ongoing support.

The Autism Inclusion Program has resulted in increased job satisfaction, improved performance, and higher retention rates among neurodivergent employees at Company X. It has created a positive and inclusive workplace culture where neurodivergent employees feel valued, supported, and empowered to reach their full potential.

Conclusion

Mentoring and coaching programs play a vital role in supporting the success and well-being of neurodivergent employees. By providing individualized support, enhancing social skills, and boosting confidence, these programs foster inclusion, unlock potential, and contribute to a more diverse and thriving workforce. Organizations must continue to invest in effective mentoring and coaching initiatives while ensuring ongoing evaluation and improvement to meet the evolving needs of neurodivergent employees in the workplace.

Managers and Supervisors as Allies

Managers and supervisors play a crucial role in creating a supportive and inclusive work environment for neurodivergent employees. By understanding their unique strengths and challenges, managers can effectively provide the necessary accommodations and support, fostering the success and well-being of their neurodivergent team members. In this section, we will explore the ways in which managers and supervisors can act as allies for their neurodivergent employees, promoting a culture of acceptance, understanding, and growth.

Understanding Neurodivergent Needs

To be effective allies, managers and supervisors need to have a comprehensive understanding of the neurodivergent conditions that their employees may have, such as autism, ADHD, or dyslexia. By educating themselves about these conditions, managers can better anticipate and address the specific needs and challenges that their neurodivergent employees may face.

It is important for managers to recognize that neurodivergent individuals may have different communication and social interaction styles. For instance, an employee with autism may prefer direct and explicit communication rather than relying on non-verbal cues. By understanding and respecting these differences, managers can create a supportive work environment that allows neurodivergent employees to thrive.

Proactive Accommodations

Managers and supervisors should take a proactive approach in identifying and implementing accommodations for their neurodivergent employees. Accommodations can include providing clear instructions, breaking down tasks into smaller steps, and offering visual aids or written instructions, which can be especially helpful for individuals with dyslexia or ADHD.

Flexible work arrangements can also support neurodivergent employees in managing their unique needs and maintaining a healthy work-life balance. This can include flexible schedules, remote work options, or adjusted workload expectations. By understanding the specific challenges faced by neurodivergent individuals, managers can tailor accommodations to support their success.

Promoting Inclusion and Sensitivity Training

To create an inclusive workplace, managers and supervisors should promote awareness and sensitivity training on neurodiversity. By organizing workshops or training sessions, managers can help employees understand the strengths and perspectives of their neurodivergent colleagues.

Inclusion training should focus on dispelling myths and stereotypes, providing practical strategies for effective communication and collaboration, and fostering a culture of empathy and understanding. By cultivating an inclusive environment, managers can create a sense of belonging and value among all employees, including those who are neurodivergent.

Supporting Career Development and Growth

Managers and supervisors can act as mentors and advocates for their neurodivergent employees, supporting their career development and growth within the organization. By providing constructive feedback, consistent support, and opportunities for skill-building, managers can help neurodivergent employees reach their full potential.

It is crucial for managers to recognize and celebrate the unique strengths that neurodivergent individuals bring to the workplace. For example, individuals with autism may have exceptional attention to detail or problem-solving skills. By recognizing and leveraging these strengths, managers can assign tasks that align with their employees' abilities and interests, leading to increased job satisfaction and productivity.

Creating a Supportive Team Environment

Managers and supervisors should foster a supportive team environment, encouraging open communication and collaboration. This can be achieved by promoting teamwork, facilitating regular team meetings, and providing opportunities for neurodivergent employees to contribute their ideas and perspectives.

It is important for managers to address any instances of discrimination or exclusion in the workplace. By cultivating a culture of mutual respect and empathy, managers can ensure that neurodivergent employees feel valued and included in the team dynamics. Additionally, managers can establish an open-door policy, where employees feel comfortable discussing any challenges or concerns they may encounter.

Case Study: Company X

Company X is a technology company that has implemented effective strategies to support their neurodivergent employees. Managers and supervisors at Company X undergo neurodiversity awareness training, which equips them with the knowledge and skills to create an inclusive work environment.

Through regular check-ins and open communication channels, managers at Company X ensure that their neurodivergent employees have the necessary support and accommodations. They also provide opportunities for skill development and growth, such as attending conferences or workshops related to their specific interests and abilities.

Company X has established a mentorship program, pairing neurodivergent employees with experienced colleagues who act as mentors and allies. This program has not only enhanced career development but has also fostered a sense of belonging and community within the organization.

By implementing these strategies, Company X has created an environment where neurodivergent employees can thrive, contributing their unique perspectives and strengths to the company's success.

Conclusion

By acting as allies, managers and supervisors play a crucial role in promoting the inclusion and success of their neurodivergent employees. By understanding their needs, providing proactive accommodations, promoting awareness and sensitivity training, supporting career development, and fostering a supportive team environment, managers can create a workplace that celebrates neurodiversity and unlocks the full potential of their workforce.

Managers and supervisors have the power to shape the culture of their organizations and create an environment where neurodivergent individuals can thrive. By embracing neurodiversity, organizations can tap into a rich pool of talents and perspectives, driving innovation, creativity, and success. It is essential for managers and supervisors to recognize their role as allies and advocates and to actively work towards a more inclusive and supportive workplace for all employees.

Disclosure and Self-Advocacy in the Workplace

In the modern workplace, neurodivergent individuals often face unique challenges related to their neurodiversity. One important aspect of navigating these challenges is disclosure - the decision to share one's neurodivergent identity with others in the workplace. Disclosing one's neurodiversity can be a personal choice, and it can have both benefits and drawbacks. This section explores the concept of disclosure and the importance of self-advocacy in the workplace for neurodivergent individuals.

What is Disclosure?

Disclosure refers to the act of sharing personal information about one's neurodivergent identity in the workplace. This could include disclosing a specific neurodevelopmental condition, such as autism or ADHD, or discussing general traits and characteristics associated with being neurodivergent. The decision to disclose is often a personal and complex one, influenced by various factors such as an individual's comfort level, the workplace culture, and potential legal protections.

Benefits of Disclosure

There are several potential benefits to disclosing one's neurodivergent identity in the workplace:

- **Accommodations and Support:** Disclosure can lead to the provision of reasonable accommodations that can help address the specific needs of neurodivergent individuals. These accommodations might include adjustments to work schedules, communication methods, or workspace arrangements. By disclosing their needs, individuals can access the support they require to thrive in their roles.

- **Reducing Misunderstandings:** Disclosure can help coworkers and supervisors better understand a neurodivergent individual's behaviors and communication style. This increased understanding can reduce misunderstandings and foster more effective collaboration and communication within the workplace.

- **Building Neurodiversity Awareness:** Disclosure can contribute to raising awareness about neurodiversity in the workplace. By openly discussing their neurodivergent identity, individuals can help educate their colleagues and promote a more inclusive workplace culture.

Drawbacks and Considerations

While disclosure can have significant benefits, it is essential to consider potential drawbacks and challenges:

- **Stigma and Discrimination:** Unfortunately, stigma and discrimination against neurodivergent individuals still exist in many workplaces. Disclosing one's neurodivergent identity may, in some cases, lead to bias, negative stereotypes, or even discrimination. These factors can create barriers to career advancement and professional growth, making the decision to disclose a complex one.

- **Privacy Concerns:** Disclosure involves sharing personal information, and some individuals may be uncomfortable with providing details about their neurodivergent identity. Privacy concerns can be particularly relevant in workplaces where there is a lack of trust, the potential for stigma, or where disclosure is not necessary for accessing accommodations.

- **Individual Autonomy:** It is crucial to respect an individual's autonomy and allow them to decide if and when to disclose their neurodivergent identity. Some individuals may feel more comfortable disclosing on a need-to-know basis, while others may choose not to disclose at all. It is essential to create a workplace environment that values and supports individual autonomy in disclosure decisions.

Self-Advocacy Strategies

Self-advocacy plays a vital role in navigating the workplace as a neurodivergent individual. By advocating for their needs and rights, neurodivergent individuals can create a supportive and inclusive work environment. Here are some self-advocacy strategies to consider:

- **Know Yourself:** Understanding one's strengths, challenges, and specific needs is crucial for effective self-advocacy. By identifying their unique abilities and requirements, neurodivergent individuals can clearly communicate their needs to coworkers and supervisors.

- **Educate Yourself and Others:** Learning about the rights and protections afforded to neurodivergent individuals in the workplace is essential. This knowledge equips individuals with the information necessary to advocate for accommodations and supports. Additionally, educating coworkers and supervisors about neurodiversity can promote understanding and inclusive practices.

- **Develop Effective Communication Skills:** Communication is key to successful self-advocacy. Neurodivergent individuals can benefit from developing strategies for effectively expressing their needs, asking for accommodations, and seeking clarification when necessary. This might involve utilizing written communication or requesting additional time to process information.

- **Seek Support and Allies:** Building a support network within the workplace can be instrumental in self-advocacy efforts. Identifying coworkers or supervisors who are understanding, accepting, and supportive can provide a valuable source of guidance and assistance in navigating workplace challenges.

- **Engage in Workplace Initiatives:** Getting involved in workplace initiatives, such as diversity and inclusion committees or employee resource groups, allows neurodivergent individuals to contribute to shaping a more inclusive work environment. This involvement not only benefits the individual but also helps foster greater awareness and understanding of neurodiversity among colleagues.

Tips for Employers and Managers

Employers and managers also play a crucial role in creating an inclusive workplace environment for neurodivergent individuals. Here are some tips for supporting disclosure and self-advocacy:

- **Promote a Culture of Acceptance and Diversity:** Employers should foster an inclusive workplace culture that values diversity, including neurodiversity. This can be achieved through education, training programs, and policies that promote acceptance and understanding.

- **Provide Clear, Confidential Channels for Disclosure:** Employers should establish clear channels for employees to disclose their neurodivergent identity, ensuring confidentiality and privacy throughout the process. This allows for open communication and ensures that appropriate accommodations can be provided.

- **Ensure Reasonable Accommodations:** Employers should work with neurodivergent individuals to provide reasonable accommodations that meet their specific needs. This might involve flexible work hours, modified job responsibilities, or workplace adjustments to support neurodivergent employees effectively.

- **Train Managers and Staff:** Training programs can help managers and staff better understand neurodiversity and develop strategies for supporting and including neurodivergent individuals. This training should focus on dispelling myths, challenging stereotypes, and promoting inclusive practices.

- **Encourage Employee Networks and Support:** Employers should support the formation of employee networks or resource groups that provide a supportive community for neurodivergent individuals. These groups can foster peer support, provide a platform for self-advocacy, and contribute to a more inclusive workplace culture.

In summary, disclosure and self-advocacy are essential aspects of navigating the workplace as a neurodivergent individual. While disclosure can have benefits such as access to accommodations and increased understanding, it is important to consider potential drawbacks and the individual's autonomy. Self-advocacy strategies assist neurodivergent individuals in effectively communicating their needs and rights, while employers and managers play a crucial role in creating an inclusive and supportive work environment. By promoting acceptance, providing accommodations, and fostering understanding, workplaces can empower neurodivergent individuals to thrive and contribute their unique talents.

Navigating Social Interactions and Communication in the Workplace

Navigating social interactions and communication in the workplace can present unique challenges for neurodivergent individuals. The ability to effectively communicate and build relationships with colleagues is essential for success in any professional setting. In this section, we will explore various strategies and techniques to help neurodivergent individuals navigate social interactions and communication in the workplace.

Understanding Social Communication

Social communication encompasses verbal and non-verbal interactions that occur in social settings. It includes the exchange of information, emotions, and intentions through spoken language, body language, facial expressions, and tone of voice. Neurodivergent individuals may face difficulties in understanding social cues, interpreting non-verbal cues, and engaging in back-and-forth conversations. Understanding the principles of social communication can help bridge these gaps and improve workplace interactions.

Recognizing and Interpreting Social Cues

One important skill in navigating social interactions is the ability to recognize and interpret social cues. Social cues can be verbal or non-verbal and may include gestures, facial expressions, tone of voice, and body language. Neurodivergent individuals may struggle with decoding these cues, leading to misunderstandings or misinterpretations. To improve social communication, it is essential to develop skills in recognizing and interpreting social cues accurately.

Example: Imagine a neurodivergent individual named Alex who has difficulty understanding sarcasm. During a team meeting, a colleague makes a sarcastic remark about a project. Alex takes the comment at face value and responds seriously. This misunderstanding can create tension and lead to miscommunication. By learning to recognize the tone of voice and facial expressions associated with sarcasm, Alex can better understand and respond appropriately in such situations.

Active Listening Skills

Active listening is a crucial component of effective communication. It involves fully engaging with the speaker, paying attention to verbal and non-verbal cues, and providing appropriate responses. Active listening can be challenging for neurodivergent individuals who may struggle with attention or processing information quickly. However, with practice, active listening skills can be developed, leading to more meaningful conversations and stronger professional relationships.

Tip: To enhance active listening skills, neurodivergent individuals can try techniques such as summarizing the speaker's key points, asking clarifying questions, and maintaining eye contact. These strategies demonstrate attentiveness and help ensure a shared understanding of the conversation.

Developing Effective Communication Strategies

To navigate social interactions successfully, neurodivergent individuals can employ various communication strategies. These strategies can help clarify their intentions, express their thoughts clearly, and facilitate smoother conversations in the workplace.

- **Use clear and concise language:** Clearly expressing thoughts and ideas can reduce the chances of miscommunication. Neurodivergent individuals can practice using simple and concise language when communicating with colleagues.

- **Ask for clarification:** If there is uncertainty or confusion during a conversation, it is important to ask for clarification. This helps to avoid misunderstandings and ensures that all parties have a shared understanding of the communication.

- **Utilize visual aids:** Visual aids, such as diagrams, charts, or infographics, can be valuable tools to enhance communication. Neurodivergent individuals can use visual aids to supplement verbal explanations and improve comprehension.

Building Relationships and Creating Support Networks

Building relationships and creating support networks in the workplace is beneficial for individuals of all neurotypes. For neurodivergent individuals, having a support network can provide a sense of belonging, understanding, and validation. It can also help navigate social challenges and provide a framework for professional growth.

Example: Connecting with colleagues who share similar experiences or interests through employee resource groups (ERGs) can be a valuable way for neurodivergent individuals to build relationships and find support in the workplace.

Addressing Social Anxiety and Stress

Social anxiety and stress can be significant barriers to effective social interactions in the workplace. Neurodivergent individuals may experience heightened anxiety in social situations or struggle with coping mechanisms to manage stress. Developing strategies to address social anxiety and stress can greatly improve overall well-being and enhance communication in the workplace.

Tip: Practicing relaxation techniques such as deep breathing, mindfulness, and self-care activities can be helpful in reducing social anxiety and stress. Taking regular breaks and engaging in activities that promote self-confidence and calmness can contribute to a more positive social experience.

Promoting Neurodiversity Acceptance in the Workplace

Creating an inclusive and accepting workplace is vital for the success and well-being of neurodivergent individuals. Employers and colleagues can play an essential role in fostering an environment that embraces neurodiversity. Educating colleagues about neurodivergent traits, strengths, and challenges can help reduce stigma and promote empathy and understanding.

Trick: Encouraging workplace activities that celebrate diversity and providing opportunities for neurodivergent individuals to share their experiences can contribute to a more inclusive and accepting workplace culture.

Summary

Navigating social interactions and communication in the workplace can be challenging for neurodivergent individuals. However, by understanding social communication, recognizing social cues, developing active listening skills, employing effective communication strategies, building relationships and support networks, addressing social anxiety and stress, and promoting neurodiversity acceptance, individuals can enhance their social interactions, foster positive relationships, and thrive in the workplace.

Keywords: social interactions, workplace communication, social cues, active listening, effective communication, relationship building, support networks, social anxiety, stress management, neurodiversity acceptance.

Mental Health and Well-being

Self-Acceptance and Neurodivergent Identity

Embracing Neurodiversity and Building Self-Esteem

In this section, we will explore the importance of embracing neurodiversity and building self-esteem for individuals who are neurodivergent. Neurodiversity refers to the natural variation in neurological development, encompassing conditions such as autism, ADHD, dyslexia, and intellectual disabilities. It emphasizes that neurodivergent individuals have unique strengths and perspectives that should be celebrated and valued in society. Building self-esteem is crucial for nurturing a positive self-image and empowering individuals to embrace their neurodivergent identities with pride.

Understanding Neurodivergent Identities

Neurodivergent individuals often face challenges related to their neurological differences. Society's neurotypical bias can lead to misunderstandings, stigma, and discrimination towards neurodivergent individuals. As a result, many neurodivergent individuals may experience feelings of inadequacy, low self-esteem, and a sense of not fitting into the neurotypical world.

To embrace neurodiversity and build self-esteem, it is essential to first understand and validate neurodivergent identities. Acknowledging that neurodivergent individuals process information, communicate, and experience the world differently is a crucial step towards promoting acceptance and inclusivity. By recognizing the unique strengths and talents associated with neurodivergent conditions, we can shift the focus from deficits to abilities.

Promoting Acceptance and Inclusion

Promoting acceptance and inclusion is fundamental to building self-esteem in neurodivergent individuals. Creating an environment that celebrates neurodiversity helps individuals feel valued, accepted, and understood. This can be achieved through various strategies:

1. Education and Awareness: Educating ourselves and others about neurodiversity fosters understanding and challenges misconceptions. By learning about the strengths and challenges associated with different neurodivergent conditions, we can promote empathy, acceptance, and reduce stigma.

2. Challenging Stereotypes: We must dismantle stereotypes and misconceptions surrounding neurodivergent individuals. Stereotypes can perpetuate a sense of shame and hinder the development of self-esteem. Highlighting positive narratives and success stories of neurodivergent individuals can help counteract these stereotypes.

3. Encouraging Self-Expression: Providing platforms for neurodivergent individuals to express themselves and share their unique perspectives is vital for building self-esteem. Art, music, writing, and other forms of creative expression can serve as powerful mediums for self-discovery, self-expression, and self-advocacy.

4. Nurturing Strengths: Recognizing and nurturing the strengths of neurodivergent individuals can boost their self-esteem. Whether it is exceptional attention to detail, creativity, or problem-solving skills, leveraging these strengths can lead to personal growth, achievement, and a sense of purpose.

Building Resilience and Coping Strategies

Neurodivergent individuals may face additional challenges and difficulties due to their neurological differences. Building resilience and developing effective coping strategies can significantly contribute to their overall well-being and self-esteem. Here are some strategies to consider:

1. Self-Acceptance: Encouraging self-acceptance involves embracing neurodivergent identities, acknowledging personal strengths, and appreciating individual differences. Developing a positive self-image promotes self-esteem and the ability to navigate challenges with resilience.

2. Seeking Support: Building a support network of understanding and empathetic individuals can provide validation, encouragement, and guidance. Support can come from friends, family, teachers, mentors, or support groups specifically tailored for neurodivergent individuals.

3. Mindfulness and self-care: Practicing mindfulness techniques can help manage stress and anxiety. Engaging in self-care activities, such as exercise, hobbies, and relaxation techniques, can improve overall well-being and foster self-esteem.

4. Setting Realistic Goals: Breaking tasks into smaller, manageable steps and setting achievable goals promotes a sense of accomplishment and self-confidence. Celebrating small victories along the way helps to reinforce self-esteem and resilience.

5. Developing Self-Advocacy Skills: Teaching self-advocacy skills can empower neurodivergent individuals to express their needs, communicate effectively, and navigate social situations. This helps individuals develop a sense of agency, autonomy, and self-worth.

Case Study: Embracing Neurodiversity and Building Self-Esteem

To illustrate the importance of embracing neurodiversity and building self-esteem, let us consider the case of Alex, a teenager with autism spectrum disorder (ASD). Alex has always struggled with social interactions and often feels disconnected from his peers. However, with the support and encouragement from his family and teachers, he discovers that he has a talent for drawing.

Alex's parents enroll him in art classes where he can explore his creativity and express himself. Through his artwork, Alex rediscovers his self-confidence and develops a sense of pride in his unique perspective. His improved self-esteem spills over into other areas of his life, including his social interactions. With newfound confidence, Alex begins to advocate for himself, expressing his needs and preferences more effectively.

Through embracing neurodiversity and nurturing his artistic talent, Alex's self-esteem flourishes. He becomes an inspiration to others, challenging stereotypes and showcasing the value of neurodivergent perspectives in the art world.

Conclusion

Embracing neurodiversity and building self-esteem are integral components of creating an inclusive and accepting society. By recognizing and valuing the unique strengths and talents of neurodivergent individuals, we can empower them to embrace their identities and navigate the world with confidence. Through education, awareness, and the promotion of resilience, we can foster an environment that celebrates neurodiversity and encourages the development of self-esteem in all individuals.

Resilience and Coping Strategies

Resilience and coping strategies play a crucial role in supporting the mental health and well-being of neurodivergent individuals. In the face of various challenges and stressors, these individuals can harness their strengths and develop effective coping mechanisms to navigate through difficult situations. This section explores the concepts of resilience and coping strategies, highlighting their importance and providing practical guidance for fostering resilience and developing coping skills.

Understanding Resilience

Resilience refers to the ability to adapt, bounce back, and maintain positive mental health despite facing adversity. It involves the capacity to cope with stress, maintain a sense of balance, and navigate through life's challenges effectively. Resilience is not a fixed trait but rather a dynamic process that can be developed and strengthened over time.

Neurodivergent individuals often face unique challenges, such as social difficulties, sensory sensitivities, and potential stigmatization. These challenges can impact their mental well-being and make resilience an essential skill to cultivate. By developing resilience, neurodivergent individuals can enhance their ability to cope with adversity, build self-esteem, and maintain positive mental health outcomes.

Factors Influencing Resilience

Several factors can influence an individual's resilience, including personal characteristics, environmental factors, and support systems. Understanding these factors can help neurodivergent individuals and those supporting them promote resilience effectively.

Personal characteristics Certain personal characteristics can contribute to an individual's resilience. These include:

- **Self-awareness:** Recognizing and understanding one's strengths, weaknesses, and emotions can help individuals better navigate challenges.
- **Self-efficacy:** Believing in one's capacity to overcome obstacles and achieve goals promotes resilience.

- **Problem-solving skills:** The ability to think critically and generate effective solutions to problems enhances resilience.

- **Positive outlook:** Maintaining a hopeful and positive perspective can bolster resilience in the face of adversity.

Environmental factors External factors in an individual's environment can significantly impact their resilience levels. Supportive environments that foster inclusion, understanding, and acceptance can contribute to resilience. Access to resources, such as educational opportunities, healthcare, and community support, can also promote resilience by providing individuals with the tools they need to navigate challenges effectively.

Support systems Having strong social support networks is crucial for developing resilience. Support systems can come from family, friends, peers, mentors, and mental health professionals. These individuals can provide emotional support, practical assistance, and guidance, which enable neurodivergent individuals to build resilience and enhance their coping strategies.

Developing Coping Strategies

Coping strategies are adaptive techniques and behaviors that individuals use to manage stress, regulate emotions, and navigate challenges effectively. Neurodivergent individuals can benefit from developing a range of coping strategies that align with their unique strengths and needs. Here are some strategies that can support their well-being:

Self-care practices Engaging in regular self-care activities can promote resilience and overall well-being. These activities may include:

- **Physical exercise:** Engaging in regular physical activity can help reduce stress, improve mood, and enhance overall mental health.

- **Stress management techniques:** Practicing deep breathing exercises, mindfulness, and relaxation techniques can help manage stress and promote emotional well-being.

- **Healthy lifestyle habits:** Maintaining a balanced diet, getting sufficient sleep, and avoiding excessive substance use can contribute to resilience.

Emotion regulation strategies Neurodivergent individuals can benefit from developing strategies to regulate their emotions, given the unique emotional experiences they may encounter. Some techniques to consider include:

- **Identifying emotions:** Developing emotional awareness and recognizing specific emotions can help individuals better understand and regulate their feelings.
- **Journaling:** Expressing emotions through written reflection can provide an outlet for processing and regulating emotions.
- **Creative outlets:** Engaging in activities such as art, music, or writing can serve as forms of self-expression and emotional release.

Seeking support Building a strong support network is critical for neurodivergent individuals. Seeking support from trusted individuals can provide validation, understanding, and practical assistance. Some strategies for seeking support may include:

- **Connecting with peers:** Building connections with others who share similar experiences can offer a sense of community and understanding.
- **Therapy and counseling:** Engaging in therapy or counseling can provide a safe space to explore emotions, develop coping strategies, and build resilience.
- **Support groups:** Participating in support groups or online communities can offer a platform for sharing experiences, gaining insights, and receiving support from others.

Promoting Neurodivergent Resilience

Promoting resilience in neurodivergent individuals requires a multi-faceted approach that integrates various strategies and support systems. Here are some key considerations:

Building self-acceptance Promoting self-acceptance is essential for fostering resilience. Encouraging neurodivergent individuals to embrace their unique strengths, talents, and qualities can enhance self-esteem and provide a solid foundation for resilience.

Educating and promoting understanding Raising awareness and promoting understanding about neurodivergent traits and experiences can foster inclusive environments that support resilience. Education about neurodiversity can reduce stigma, encourage empathy, and cultivate acceptance.

Collaboration and advocacy Collaboration among educators, parents, mental health professionals, and the neurodivergent individuals themselves is crucial for promoting resilience. Advocacy efforts should focus on creating supportive policies, inclusive educational settings, and accessible resources to facilitate the development of resilience in neurodivergent individuals.

Strength-based approaches Emphasizing and building on the strengths of neurodivergent individuals can enhance resilience. By recognizing and developing their unique talents and abilities, individuals can build self-confidence and a positive outlook, which are essential components of resilience.

Celebrating diversity Promoting a culture that celebrates diversity and embraces neurodivergent perspectives can contribute to resilience. Valuing the contributions and perspectives of neurodivergent individuals helps create an inclusive environment that fosters growth, resilience, and overall well-being.

Case Study: Building Resilience in an Autistic Student

Consider the case of Emily, a high school student diagnosed with Autism Spectrum Disorder (ASD). Emily faces challenges related to sensory sensitivities, social interactions, and transitioning between tasks. To support her resilience, her teachers and support team implement the following strategies:

- **Sensory accommodations:** Providing a quiet space for breaks and allowing the use of sensory tools, such as noise-cancelling headphones or fidget toys, helps Emily regulate her sensory experiences.

- **Social skills training:** Engaging in social skills training sessions equips Emily with the tools and strategies necessary for navigating social interactions successfully.

- **Collaborative planning:** Involving Emily in the development of her Individualized Education Plan (IEP) allows her to have a voice in her education and builds self-advocacy skills.

- **Strength-based approach:** Recognizing and celebrating Emily's strengths, such as her attention to detail and creativity, boosts her self-esteem and resilience.

- **Mentoring support:** Pairing Emily with a mentor who has experience with ASD helps provide guidance, support, and a role model who can offer strategies for overcoming challenges.

By implementing these strategies and supporting Emily's unique needs, her resilience is nurtured, enabling her to navigate the school environment successfully and thrive academically and socially.

Conclusion

Resilience and coping strategies are integral to supporting the well-being of neurodivergent individuals. By understanding the factors that influence resilience, individuals and support systems can promote the development of effective coping strategies. Through self-care practices, emotion regulation techniques, seeking support, and promoting inclusive environments, neurodivergent individuals can build resilience, navigate challenges, and thrive in a world that celebrates and embraces their differences.

Neurodivergent Narratives and Autobiographical Accounts

Neurodivergent narratives and autobiographical accounts are powerful tools for understanding and embracing neurodiversity. These stories, shared by individuals who are neurodivergent themselves, provide unique insights into their experiences, challenges, strengths, and perspectives. By amplifying these voices, we gain a deeper understanding of what it means to be neurodivergent and foster empathy, acceptance, and inclusivity in our society.

1. The Power of Personal Stories:

Autobiographical narratives allow individuals to share their personal journeys of self-discovery and self-acceptance. They provide a platform to challenge stereotypes and misconceptions surrounding neurodivergence. These stories help in dispelling the notion that neurodivergent individuals are broken or in need of fixing. Instead, they highlight the diverse talents, skills, and perspectives that neurodivergent individuals bring to the table.

2. Promoting Empathy and Understanding:

Neurodivergent narratives cultivate empathy and understanding among neurotypical individuals. By reading or listening to the stories of neurodivergent individuals, we not only gain insight into their unique experiences but also learn to appreciate the strengths and challenges they face. This understanding breaks down barriers and fosters a more inclusive and accepting society.

3. Overcoming Stigma and Stereotypes:

Neurodivergent narratives challenge stereotypes and break down the stigma associated with neurodivergence. By sharing their personal stories, individuals highlight their achievements, talents, and passions. These narratives help in showcasing the diversity within the neurodivergent community and challenging limited perceptions of what it means to be neurodivergent.

4. Providing Role Models and Inspiration:

Neurodivergent autobiographies serve as a source of inspiration and strength for individuals who are on their own neurodivergent journey. These narratives provide role models who have navigated the challenges and successes of being neurodivergent and offer hope to those who may be struggling. By sharing their stories, neurodivergent individuals inspire others to embrace their differences and strive for their goals.

5. Amplifying Lesser Heard Voices:

Neurodivergent narratives ensure that the voices of marginalized neurodivergent individuals are heard. These stories bring attention to the experiences of underrepresented groups within the neurodivergent community, such as individuals from minority backgrounds or individuals with intersecting identities. By amplifying these voices, we create a more inclusive and diverse conversation around neurodiversity.

6. Resources for Education and Awareness:

Autobiographical accounts provide valuable resources for education and awareness about neurodiversity. These stories can be used in educational settings to foster understanding and acceptance among students. They can also be shared in workplaces, community groups, and other settings to promote inclusive practices and challenge discriminatory attitudes.

Example Story: "Finding My Voice"

In "Finding My Voice," Sarah shares her journey of self-discovery as an autistic woman. Growing up, Sarah struggled to understand her differences and often felt like an outsider. With the support of her family and friends, she embarked on a journey of self-acceptance, embracing her neurodivergent identity.

Sarah's story highlights the challenges she faced in navigating social situations and sensory sensitivities. Through her determination and resilience, she found her passion for writing and advocacy. Sarah's journey resonates with readers who are navigating their own neurodivergent paths, providing them with hope and inspiration.

Exercise: Reflection and Journaling

Take a moment to reflect on your own experiences and perceptions of neurodivergent individuals. How have personal stories and narratives shaped your understanding and empathy? Journal about an encounter or experience that challenged your preconceived notions about neurodiversity and inspired greater acceptance.

By embracing and amplifying neurodivergent narratives and autobiographical accounts, we can create a more inclusive and understanding society. These stories challenge stereotypes, break down barriers, and inspire others to embrace their own neurodivergent identities. Let us celebrate the unique voices and experiences of neurodivergent individuals, fostering acceptance, and promoting neurodiversity in all aspects of life.

The Influence of Neurodiversity on Personal Relationships

Neurodiversity, the recognition and acceptance of neurological differences, has a significant impact on personal relationships. Understanding how neurodivergent individuals experience and navigate social connections is crucial for fostering meaningful and inclusive relationships. In this section, we will explore the ways in which neurodiversity influences personal relationships, the challenges that may arise, and strategies for promoting understanding and communication.

The Neurodivergent Experience in Relationships

Neurodivergent individuals often bring unique perspectives and strengths to their relationships. Their different cognitive styles, processing preferences, and sensory sensitivities can contribute to diverse ways of thinking, problem-solving, and experiencing the world. This can enrich relationships by fostering creativity, innovation, and alternative approaches to challenges.

At the same time, neurodivergent individuals may face particular challenges in social interactions and emotional connections. Differences in social communication, such as difficulties with nonverbal cues or challenges in understanding sarcasm or metaphor, can make it harder for neurodivergent individuals to interpret social signals and navigate social expectations. This can sometimes lead to misunderstandings, frustration, and feelings of isolation.

Moreover, the impact of neurodivergent conditions on executive functioning, emotional regulation, and sensory processing can affect how individuals navigate emotional intimacy, conflict resolution, and shared activities within relationships.

Understanding and addressing these challenges is crucial for fostering healthy and fulfilling relationships.

Challenges in Neurodivergent Relationships

Neurodivergent individuals may face unique challenges in forming and sustaining meaningful relationships. Some common challenges include:

- **Communication Differences:** Neurodivergent individuals may struggle with social cues, indirect communication, or maintaining reciprocal conversations. This can lead to misinterpretations, conflicts, or feelings of disconnection.
- **Sensory Sensitivities:** Many individuals with neurodivergent conditions experience sensory sensitivities that can impact their tolerance for certain environments or activities. This can create barriers to shared experiences or limit opportunities for connection.
- **Executive Functioning:** Difficulties with planning, organizing, and managing time can affect the ability to prioritize and meet relationship commitments, leading to misunderstandings or strain.
- **Emotional Regulation:** Neurodivergent individuals may experience challenges in recognizing and regulating their emotions, leading to difficulty expressing needs or navigating conflicts.
- **Social Anxiety:** The fear of judgment or rejection can be heightened in neurodivergent individuals, potentially impacting their ability to initiate or engage in social interactions.

It is important to recognize that these challenges are not inherent to neurodivergent individuals, but rather arise due to a lack of understanding, acceptance, and support from neurotypical society.

Strategies for Building Stronger Relationships

Building and nurturing relationships with neurodivergent individuals can be a rewarding experience. Here are some strategies to promote understanding, communication, and connection:

- **Education and Awareness:** Educate yourself about neurodivergent conditions and their impact on relationships. This will enable you to better understand and appreciate the experiences and perspectives of neurodivergent individuals.

- **Open Communication:** Foster an environment of open and direct communication, where both parties can express their needs, preferences, and concerns. Clarify communication styles and develop strategies for effective communication.
- **Respect and Acceptance:** Embrace neurodiversity and appreciate the unique strengths and differences that neurodivergent individuals bring to the relationship. Avoid judgment and prioritize acceptance.
- **Accommodations and Flexibility:** Make accommodations to create a supportive and inclusive environment that respects the specific needs and preferences of neurodivergent individuals. Be flexible in adapting to different communication styles and sensory sensitivities.
- **Emotional Support:** Recognize and validate the emotions of neurodivergent individuals. Be patient, empathetic, and non-judgmental in providing emotional support during challenging situations.
- **Shared Activities and Interests:** Engage in activities or interests that provide a common ground and promote shared experiences. Respect individual boundaries and preferences while finding opportunities for connection.
- **Seeking Professional Support:** Consider seeking the guidance of therapists, counselors, or support groups that specialize in neurodiversity and relationships. They can provide valuable insights and strategies tailored to your specific situation.

Remember, building and maintaining relationships is a dynamic process that requires effort, understanding, and empathy. By embracing neurodiversity, fostering open communication, and celebrating individual strengths, we can create supportive and enriching relationships that celebrate our differences.

Case Study: The Power of Understanding

Emily and Peter have been in a romantic relationship for several years. Peter is neurodivergent and has been diagnosed with ADHD, while Emily is neurotypical. Early in their relationship, they encountered challenges related to communication and emotional understanding, which strained their connection.

Through open and honest conversations, they began to learn about each other's perspectives and differences. Emily made an effort to understand the impact of ADHD on Peter's executive functioning and worked collaboratively to create strategies to manage daily tasks and responsibilities. Peter, in turn, explained his preferred communication style and sensitivities, enabling Emily to adapt and better meet his needs.

By embracing neurodiversity and committing to open dialogue, Emily and Peter strengthened their relationship. They celebrated their unique strengths and valued the different perspectives they brought to the table. Their relationship flourished as they developed a deep understanding and appreciation for each other.

Conclusion

Neurodiversity plays a significant role in personal relationships. Recognizing and celebrating neurodivergent traits can lead to more inclusive and fulfilling connections. By promoting understanding, communication, and empathy, we can build stronger relationships that value and appreciate the unique contributions of neurodivergent individuals.

Navigating Social Challenges and Building Support Networks

Navigating social challenges and building support networks can be particularly important for neurodivergent individuals who often face difficulties in social interactions. In this section, we will explore some of the common social challenges faced by neurodivergent individuals and discuss strategies for building support networks that can provide the necessary guidance and understanding.

Social Challenges Faced by Neurodivergent Individuals

Neurodivergent individuals may face a range of social challenges due to differences in communication styles, sensory sensitivities, and social cognition. Some of the common social challenges include:

- Difficulty understanding social cues and non-verbal communication
- Challenges in initiating and maintaining conversations
- Sensory overload in social situations, leading to anxiety or withdrawal
- Misinterpretation of social norms and expectations

- Experiences of social exclusion and bullying

These challenges can often make it challenging for neurodivergent individuals to form and maintain meaningful social relationships. However, with the right support and strategies, they can overcome these challenges and build a strong support network.

Strategies for Navigating Social Challenges

Here are some strategies that can help neurodivergent individuals navigate social challenges and build support networks:

1. **Education and self-awareness:** It is essential for neurodivergent individuals to understand their own strengths, challenges, and communication styles. By learning about their neurodivergent traits, they can gain self-confidence and communicate their needs effectively to others.

2. **Social skills training:** Engaging in social skills training programs can help neurodivergent individuals learn and practice social skills in a structured environment. These programs often provide guidance on non-verbal communication, active listening, and understanding social cues.

3. **Joining support groups:** Participating in support groups specifically tailored for neurodivergent individuals can provide a safe space to share experiences, challenges, and strategies. Support groups can offer opportunities for social interactions, where individuals can learn from each other and build supportive relationships.

4. **Networking events and organizations:** Attending networking events and joining organizations focused on neurodiversity can help neurodivergent individuals connect with like-minded individuals and professionals who understand their unique challenges. These communities can offer support, advice, and mentorship opportunities.

5. **Peer mentorship:** Engaging in peer mentorship programs, where neurodivergent individuals are paired with more experienced mentors, can be beneficial. Mentors can provide guidance in navigating social challenges, sharing strategies based on their own experiences, and offering support.

6. **Online communities:** Online platforms and social media groups dedicated to neurodiversity can be valuable resources for neurodivergent individuals.

These communities allow individuals to connect and share their experiences, seek advice, and gain support from people who understand their journey.

7. **Building a support network:** It is important for neurodivergent individuals to build a strong support network of friends, family, educators, and professionals who can provide guidance, understanding, and encouragement. These individuals can serve as a sounding board, offer perspective, and help navigate challenging social situations.

8. **Advocacy and self-advocacy:** Becoming an advocate for neurodiversity and educating others about the challenges faced by the community can help foster understanding and acceptance. Additionally, learning effective self-advocacy skills can empower neurodivergent individuals to communicate their needs, rights, and boundaries.

Real-World Example: Autism Self-Advocacy Network

The Autism Self-Advocacy Network (ASAN) is an example of an organization that focuses on empowering autistic individuals and promoting self-advocacy. ASAN provides resources, training, and advocacy initiatives to help autistic individuals navigate social challenges, build support networks, and advocate for their rights.

ASAN offers mentorship programs, support groups, and online communities where autistic individuals can connect with others, share their experiences, and learn from one another. They also conduct workshops and training sessions to enhance social skills, self-advocacy, and overall well-being.

Through their advocacy efforts, ASAN works towards creating a more inclusive society by raising awareness about the strengths and challenges of autism and promoting acceptance and inclusion. They collaborate with stakeholders, policymakers, and educators to ensure that the rights and needs of autistic individuals are recognized and respected.

Takeaway

Navigating social challenges and building support networks are critical components of fostering social inclusion and well-being for neurodivergent individuals. By equipping themselves with strategies, education, and advocacy tools, neurodivergent individuals can develop the skills and networks necessary to overcome social obstacles and thrive in a neurotypical world. It is important to recognize the value of diversity and create an inclusive society where everyone's strengths and contributions are celebrated.

Mental Health Concerns in Neurodivergent Individuals

Anxiety and Depression in Neurodivergent Individuals

Anxiety and depression are common mental health concerns that can affect individuals across the neurodivergent spectrum. In this section, we will explore the unique challenges faced by neurodivergent individuals in relation to anxiety and depression, as well as strategies for supporting their mental well-being.

Understanding the Relationship between Neurodivergence and Anxiety/Depression

Neurodivergent individuals may experience anxiety and depression as a result of various factors related to their neurology and the social environment. It is important to recognize that anxiety and depression can co-occur with other neurodevelopmental conditions, such as autism spectrum disorder (ASD), attention-deficit/hyperactivity disorder (ADHD), and dyslexia.

One of the contributing factors to anxiety and depression in neurodivergent individuals is the heightened sensory and emotional sensitivity often associated with their neurology. For example, individuals with ASD may experience sensory overload in certain environments, leading to heightened anxiety. Additionally, difficulties with social communication and interaction can contribute to feelings of isolation and loneliness, which can further exacerbate depression.

Another factor is the impact of stigma and discrimination. Neurodivergent individuals may face challenges in accessing appropriate support services, accommodations, and acceptance in various social settings. These experiences can lead to feelings of inadequacy, rejection, and low self-esteem, increasing the risk of anxiety and depression.

Recognizing the Signs and Symptoms

It is crucial to be aware of the signs and symptoms of anxiety and depression in neurodivergent individuals, as they may present differently compared to neurotypical individuals. The following are some common indicators:

- **Anxiety**

  - Heightened sensitivity to sensory stimuli (e.g., noise, lights)
  - Avoidance of certain situations or places that trigger anxiety
  - Excessive worry or rumination
  - Physical symptoms, such as rapid heartbeat, difficulty breathing, and gastrointestinal distress
  - Restlessness or agitation
  - Sleep disturbances
  - Difficulty with transitions or changes in routine

- **Depression**
  - Persistent feelings of sadness or hopelessness
  - Loss of interest or pleasure in previously enjoyed activities
  - Changes in appetite or sleep patterns
  - Fatigue and decreased energy levels
  - Difficulty concentrating or making decisions
  - Withdrawal from social interactions
  - Thoughts of self-harm or suicide (this is an emergency, and immediate professional help should be sought)

Supporting Neurodivergent Individuals with Anxiety and Depression

Creating a supportive environment and providing appropriate interventions can significantly improve the well-being of neurodivergent individuals experiencing anxiety and depression. Here are some strategies for supporting their mental health:

- **Early Intervention:** Early identification and intervention are crucial. Educators, family members, and healthcare professionals should be vigilant in recognizing signs of anxiety and depression to ensure timely intervention.

- **Person-Centered Approaches:** Taking an individualized approach is essential, considering the unique needs and preferences of neurodivergent individuals. Collaborate with them to develop strategies that work best for them.

- **Sensory Regulation:** Provide opportunities for sensory regulation by creating calm and quiet spaces, allowing the use of noise-canceling headphones or other sensory tools, and minimizing sensory triggers.

- **Social Skills Training:** Offer social skills training programs to enhance social communication and interaction skills, reducing feelings of isolation and anxiety.

- **Cognitive-Behavioral Therapy (CBT):** CBT can be beneficial for managing anxiety and depression. Adapting CBT techniques to accommodate the cognitive and sensory profiles of neurodivergent individuals may be necessary.

- **Supportive School and Work Environments:** Educators and employers should create inclusive environments that offer appropriate accommodations, establish clear expectations, and promote understanding and acceptance.

- **Peer Support and Mentoring Programs:** Facilitate peer support programs where neurodivergent individuals can connect with others facing similar challenges and receive guidance from mentors who have overcome similar struggles.

- **Multidisciplinary Collaboration:** Foster collaboration among educators, healthcare professionals, therapists, and families to develop comprehensive support plans that address both mental health and academic/work-related needs.

- **Recognize Strengths and Celebrate Progress:** Focus on celebrating the strengths and achievements of neurodivergent individuals to foster a positive self-image and build resilience.

It is important to note that supporting neurodivergent individuals with anxiety and depression requires a holistic approach and a commitment to promoting acceptance, inclusion, and understanding. By providing the necessary support, we can help individuals thrive and achieve their full potential. Remember, seeking professional advice from mental health professionals is essential for accurate diagnosis and personalized interventions.

Real-World Example: A School's Approach

Riverdale High School has implemented a comprehensive program to support neurodivergent students with anxiety and depression. The program includes the following components:

- **Mental Health Awareness Workshops:** Regular workshops are conducted for staff and students to increase awareness and understanding of anxiety and depression in neurodivergent individuals.
- **Student Support Teams:** The school has formed multidisciplinary support teams consisting of teachers, counselors, and therapists who collaborate to develop individualized support plans for students with anxiety and depression.
- **Sensory-Friendly Spaces:** Several sensory-friendly spaces have been created throughout the school, providing a calming environment for students to regulate their sensory needs.
- **Social Skills Training:** Regular social skills training sessions are offered to help neurodivergent students enhance their social communication and interpersonal skills.
- **Peer Mentoring Program:** A peer mentoring program pairs neurotypical students with neurodivergent students, fostering supportive relationships and easing feelings of social isolation.
- **Teacher Training:** The school provides ongoing training for teachers to enhance their understanding of anxiety and depression in neurodivergent individuals and equips them with strategies to support students effectively.

Riverdale High School's program has led to improved mental well-being, academic performance, and overall school experience for its neurodivergent students. Other schools can take inspiration from their approach when designing their own support programs.

Resources and Further Reading

- "The Anxiety and Depression Workbook for Neurodivergent Individuals" by Emma Goodall
- "All Birds Have Anxiety" by Kathy Hoopmann

- "Uniquely Human: A Different Way of Seeing Autism" by Barry M. Prizant
- "Neurodiversity in the Classroom: Strength-Based Strategies to Help Students with Special Needs Succeed in School and Life" by Thomas Armstrong
- "Overcoming Depression and Anxiety for Autism and Neurodiversity: A Self-Help Guide Using CBT" by Lee A. Wilkinson
- "Mindfulness for Beginners" by Jon Kabat-Zinn

Internet resources:

- Autistic Self Advocacy Network (ASAN) - `https://autisticadvocacy.org/`
- Anxiety and Depression Association of America (ADAA) - `https://adaa.org/`

Exercise: Reflect on the strategies mentioned in this section and think of additional measures that can support neurodivergent individuals with anxiety and depression. Share your ideas with a classmate or write them down in a journal.

Remember, everyone's experience with anxiety and depression is unique, and it is essential to tailor interventions and support to each individual's specific needs. By promoting understanding, acceptance, and access to appropriate resources, we can create a more inclusive and supportive environment for neurodivergent individuals grappling with anxiety and depression.

Supporting Individuals with Co-occurring Mental Health Conditions

When it comes to neurodivergent individuals, it is important to recognize that many of them may also experience co-occurring mental health conditions. Co-occurring mental health conditions refer to the presence of two or more mental health disorders in an individual. This intersection of neurodivergence and mental health requires specialized support and understanding to ensure the well-being of these individuals. In this section, we will explore strategies and resources for supporting individuals with co-occurring mental health conditions within the framework of neurodiversity.

Understanding Co-occurring Mental Health Conditions

To effectively support individuals with co-occurring mental health conditions, it is essential to have a comprehensive understanding of the various disorders and their interactions. Some common co-occurring mental health conditions among neurodivergent individuals include anxiety disorders, depressive disorders, obsessive-compulsive disorder (OCD), bipolar disorder, and schizophrenia.

Person-Centered Approach to Support

A person-centered approach is crucial when supporting individuals with co-occurring mental health conditions. This approach acknowledges and respects an individual's unique experiences and needs, promoting their active involvement in decision-making and goal-setting. It is essential to recognize that every individual's experience of co-occurring mental health conditions is different, and tailor support accordingly.

Collaboration with Mental Health Professionals

Collaboration with mental health professionals, including psychologists, psychiatrists, and therapists, is vital in supporting individuals with co-occurring mental health conditions. These professionals can provide expertise in diagnosis, treatment, and ongoing support. By working together, a collaborative care model can be established to ensure comprehensive and holistic support for neurodivergent individuals with co-occurring mental health conditions.

Integrated Treatment Approaches

Integrated treatment approaches are beneficial for individuals with co-occurring mental health conditions, as they address both neurodivergent traits and mental health concerns simultaneously. These approaches involve the coordination of services and treatment plans from both neurodiversity and mental health perspectives. An integrated treatment plan may include a combination of medications, therapy (such as cognitive-behavioral therapy or dialectical behavior therapy), social support, and skill-building interventions.

Creating Supportive Environments

Supportive environments play a crucial role in promoting the well-being of individuals with co-occurring mental health conditions. These environments should foster acceptance, understanding, and inclusion, ensuring that individuals feel safe and supported. Education and training for family members, friends, teachers, and colleagues can help create a supportive network for the individual. Additionally, practicing sensitivity, empathy, and patience are essential when interacting with individuals who may be navigating the challenges of both neurodivergence and mental health conditions.

Self-Care and Coping Strategies

Teaching individuals with co-occurring mental health conditions effective self-care and coping strategies is vital for their overall well-being. This may include techniques for stress management, mindfulness exercises, relaxation techniques, and developing healthy routines. Encouraging individuals to engage in activities they enjoy, maintain a balanced lifestyle, and seek support from their social network can contribute to their mental health resilience.

Peer Support and Community Resources

Peer support can be instrumental in the journey of individuals with co-occurring mental health conditions. Connecting with others who have similar experiences can provide a sense of belonging, understanding, and validation. Peer support groups and online communities specifically designed for neurodivergent individuals with co-occurring mental health conditions can serve as a valuable resource.

Advocacy and Policy Changes

Advocacy plays a crucial role in promoting change and ensuring adequate support for individuals with co-occurring mental health conditions. Advocacy efforts can focus on raising awareness, challenging stigma, and advocating for policy changes that prioritize the needs and rights of neurodivergent individuals. It is important to involve neurodivergent individuals themselves in these advocacy efforts, empowering them to share their experiences and actively participate in shaping policies and services.

In conclusion, supporting individuals with co-occurring mental health conditions requires taking a person-centered approach, collaborating with mental health professionals, implementing integrated treatment approaches, creating supportive environments, teaching self-care and coping strategies, utilizing peer support and community resources, as well as advocating for policy changes. By adopting these strategies within the framework of neurodiversity, we can ensure the well-being and empowerment of individuals who are navigating the intersection of neurodivergence and mental health conditions.

Trauma and Neurodivergent Experiences

Trauma refers to an emotional or psychological response to a distressing or disturbing event. It can have a profound and long-lasting impact on an individual's mental well-being. Neurodivergent individuals, who already navigate unique challenges related to their neurodevelopmental conditions, may experience trauma in different ways than their neurotypical peers. In this section, we will explore the intersection between trauma and neurodivergent experiences, examining the potential consequences and highlighting strategies to support and promote healing.

Understanding Trauma in Neurodivergent Individuals

Neurodivergent individuals often experience heightened sensory sensitivity and difficulty with social communication, which can contribute to increased vulnerability to traumatic experiences. For example, individuals with autism spectrum disorder (ASD) may have difficulties discerning social cues, leading to misunderstandings, exclusion, and potential maltreatment. Those with ADHD may struggle with impulse control, which can put them at risk for impulsive, risky behaviors and subsequently traumatic events.

It is essential to recognize that trauma can manifest differently in neurodivergent individuals compared to neurotypical individuals. Some common reactions to trauma in neurodivergent individuals may include increased anxiety, difficulty with emotional regulation, sleep disturbances, and heightened sensory sensitivities. However, these reactions can be masked or misinterpreted, making it crucial for caregivers, educators, and healthcare professionals to have a deeper understanding of how trauma may manifest in this population.

Addressing Trauma in Neurodivergent Individuals

Supporting neurodivergent individuals who have experienced trauma requires a comprehensive and individualized approach. Here are several strategies that can be employed:

1. **Trauma-informed care:** Adopting a trauma-informed approach involves creating a safe and supportive environment that acknowledges and responds to the specific needs of neurodivergent individuals who have experienced trauma. This

approach emphasizes building trust, promoting empowerment, and recognizing the interconnectedness between trauma and neurodivergent experiences.

2. **Sensory integration techniques:** Sensory integration techniques can be beneficial for neurodivergent individuals who have experienced trauma. These techniques involve engaging the senses in a structured and therapeutic manner, aimed at regulating sensory responses and promoting emotional well-being. Activities such as deep pressure touch, rhythmic movements, and sensory diet interventions can help individuals regulate their emotions and regain a sense of security.

3. **Social and emotional skill development:** Teaching social and emotional skills is crucial for neurodivergent individuals who have experienced trauma. This includes providing explicit instruction in recognizing emotions, coping strategies for emotional regulation, problem-solving, and assertiveness training. By developing these skills, individuals can improve their ability to navigate social relationships and cope with trauma-related triggers.

4. **Therapeutic interventions:** Various therapeutic interventions can be effective in addressing trauma in neurodivergent individuals. For example, cognitive-behavioral therapy (CBT) can help individuals identify and challenge negative thought patterns and develop effective coping mechanisms. Eye movement desensitization and reprocessing (EMDR) is another therapeutic approach that has shown promise in treating trauma-related symptoms.

Case Study: Jane's Journey to Healing

To illustrate the impact of trauma on neurodivergent individuals and the importance of trauma-informed care, let's consider Jane, a 14-year-old girl with ADHD. Jane experienced bullying at school, leading to feelings of isolation, low self-esteem, and anxiety. These experiences left a lasting impact on her emotional well-being and academic performance.

Jane's school recognized the need for a trauma-informed approach and implemented the following strategies:

- Provide a safe and supportive environment where Jane feels comfortable expressing her emotions and discussing her experiences.
- Offer individual counseling sessions with a trauma-informed therapist who specializes in working with neurodivergent individuals.

- Teach Jane coping strategies for emotional regulation, such as deep-breathing exercises and mindfulness techniques.
- Collaborate with Jane's teachers to create an individualized education plan (IEP) that addresses her specific needs, including accommodations for sensory sensitivities and additional support for academic challenges.
- Foster a sense of belonging and social connectedness by encouraging participation in extracurricular activities that align with Jane's interests.

Over time, with consistent support and trauma-informed care, Jane began to heal from her traumatic experiences. Her anxiety diminished, and her self-confidence improved, allowing her to thrive both academically and socially.

Conclusion

Trauma can profoundly impact the lives of neurodivergent individuals, exacerbating the challenges they already navigate due to their neurodevelopmental conditions. Recognizing the unique intersection between trauma and neurodivergent experiences is essential for providing effective support and promoting healing.

By adopting a trauma-informed approach, employing sensory integration techniques, investing in social and emotional skill development, and utilizing therapeutic interventions, caregivers, educators, and health professionals can create an environment that fosters resilience and growth for neurodivergent individuals who have experienced trauma. With targeted support, these individuals can regain a sense of safety, build healthy coping mechanisms, and thrive in their personal and academic pursuits.

The Role of Stress Management and Mindfulness

Stress is a common experience for everyone, but it can have a particularly significant impact on neurodivergent individuals. The daily challenges and sensory sensitivities they face can contribute to higher stress levels, making it crucial to develop effective stress management techniques. One approach that has gained significant attention in recent years is mindfulness.

Understanding Stress and Its Effects

Before delving into stress management techniques, it's important to understand what stress is and how it affects neurodivergent individuals. Stress is the body's response to a perceived threat or challenge, triggering a cascade of physiological and psychological reactions. In the neurodivergent population, stress can be triggered by various factors, such as sensory overload, social interactions, and changes in routine.

The effects of stress on neurodivergent individuals can be significant. It can lead to increased anxiety, meltdowns, difficulties with attention and focus, and compromised mental well-being. Therefore, incorporating stress management techniques into their daily lives is crucial for ensuring their overall well-being.

Introducing Mindfulness

Mindfulness is a practice that involves paying attention to the present moment without judgment. It originated from Buddhist traditions and has been adapted for secular purposes, including stress reduction and overall well-being. Mindfulness can be practiced through various activities such as meditation, deep breathing exercises, or simply being fully present in the moment.

For neurodivergent individuals, mindfulness can be a powerful tool for managing stress. By practicing mindfulness, they can develop greater self-awareness, enhance their ability to regulate emotions, and cultivate a sense of calmness and resilience in the face of challenges.

Benefits of Mindfulness for Neurodivergent Individuals

1. Stress Reduction: Mindfulness helps neurodivergent individuals reduce stress by shifting their focus away from anxiety-inducing thoughts and sensations. By redirecting their attention to the present moment, they can alleviate stress and promote relaxation.

2. Emotional Regulation: Neurodivergent individuals often experience difficulties in regulating their emotions. Mindfulness practices, such as mindful breathing, can help them become more attuned to their emotional states and develop strategies for effectively managing and expressing emotions.

3. Improved Attention and Focus: Attention deficits are common among neurodivergent individuals. Mindfulness practices, such as mindful meditation, can enhance their ability to sustain attention and improve overall focus.

4. Increased Self-Awareness: Mindfulness cultivates self-awareness, allowing neurodivergent individuals to develop a deep understanding of their own needs, triggers, and strengths. This self-awareness can empower them to make self-care choices that support their well-being.

5. Resilience and Coping Skills: Neurodivergent individuals often face unique challenges that require resilience and adaptive coping skills. Mindfulness practices can enhance their ability to bounce back from difficulties, reduce rumination, and improve problem-solving skills.

Incorporating Mindfulness into Daily Life

Here are some practical strategies for incorporating mindfulness into the daily lives of neurodivergent individuals:

1. Mindful Breathing: Encourage individuals to take a few moments each day to focus on their breath. They can practice deep, slow breaths and notice the sensations in their body as they inhale and exhale.

2. Mindful Body Scan: Guide individuals through a body scan meditation, where they systematically bring attention to each part of their body, noticing any tension or discomfort and allowing it to release.

3. Mindful Movement: Engage individuals in activities such as yoga, qigong, or gentle stretching, encouraging them to pay attention to their body's movements, sensations, and the present moment.

4. Mindful Walking: Encourage individuals to take mindful walks, paying attention to the sensations in their feet, the sights, sounds, and smells around them, and the act of walking itself.

5. Mindful Listening: Guide individuals through a mindful listening exercise, where they focus their attention on sounds in their environment without judgment or analysis.

6. Mindful Journaling: Encourage individuals to keep a mindfulness journal, where they can reflect on their daily experiences and feelings, practicing mindful awareness as they write.

It's important to remember that mindfulness is a skill that requires practice and patience. Neurodivergent individuals may find it helpful to start with shorter mindfulness sessions and gradually increase the duration as they become more comfortable with the practice.

The Importance of Individualization

When introducing mindfulness to neurodivergent individuals, it's essential to consider their unique needs, preferences, and sensory profiles. As the neurodivergent community is diverse, there is no one-size-fits-all approach to mindfulness. Some individuals may respond well to certain practices, while others may find them overwhelming or ineffective.

Therefore, it is crucial to provide a range of options and encourage individuals to explore what works best for them. Mindfulness should always be presented as a tool that individuals can adapt and personalize according to their specific needs and preferences.

A Word of Caution

While mindfulness can be highly beneficial for many neurodivergent individuals, it is essential to acknowledge that it may not be suitable or effective for everyone. Some individuals may find certain mindfulness practices uncomfortable or triggering due to their sensory sensitivities or past traumas. Therefore, it is crucial to provide alternative stress management techniques and support individuals in finding strategies that work best for them.

Conclusion

Incorporating stress management techniques, such as mindfulness, can be immensely valuable for neurodivergent individuals. By cultivating self-awareness, emotional regulation, and resilience, they can navigate the challenges of daily life with greater ease and well-being. However, it is crucial to approach mindfulness with a person-centered and individualized mindset, considering the unique needs and preferences of each individual. Mindfulness should be seen as one tool among many, with flexibility and adaptability as guiding principles in supporting neurodivergent individuals in their stress management journey.

Accessible Therapeutic Approaches for Neurodivergent Individuals

Accessible therapeutic approaches for neurodivergent individuals are crucial in promoting their mental health and well-being. These approaches aim to address the unique challenges faced by neurodivergent individuals while considering their individual strengths and needs. In this section, we will explore some of the key therapeutic approaches that have been effective in supporting neurodivergent individuals.

Applied Behavior Analysis (ABA)

Applied Behavior Analysis (ABA) is a therapeutic approach that focuses on understanding and modifying behaviors in individuals with neurodevelopmental conditions such as autism. ABA uses evidence-based techniques to teach new skills, decrease problematic behaviors, and improve social interactions.

ABA therapy uses the principles of behaviorism to target specific behaviors and shape them through positive reinforcement. The therapist sets clear goals and breaks down complex tasks into smaller, more manageable steps to facilitate learning. This approach emphasizes the use of visual supports, structured environments, and repetition to enhance comprehension and generalization of skills.

One example of ABA therapy is discrete trial training, where the therapist uses repetition and reinforcement to teach specific skills. In this method, the individual is presented with a clear instruction, completes the task, and receives immediate feedback and reinforcement. This process is repeated to reinforce learning and build fluency.

While ABA can be effective, it is important to ensure that it is implemented ethically and respects the rights and dignity of the individual. Critics argue that ABA should focus on promoting autonomy and self-determination rather than simply modifying behaviors.

Cognitive-Behavioral Therapy (CBT)

Cognitive-Behavioral Therapy (CBT) is a widely used therapeutic approach that focuses on the connection between thoughts, feelings, and behaviors. CBT aims to help individuals identify and challenge negative thoughts and beliefs, develop coping strategies, and improve emotional regulation.

CBT can be adapted to meet the unique needs of neurodivergent individuals by incorporating visual supports, concrete examples, and structured approaches. It can be particularly beneficial for individuals with anxiety, depression, or other co-occurring mental health conditions.

One technique commonly used in CBT is cognitive restructuring, where individuals are encouraged to challenge and reframe negative thoughts. For example, a neurodivergent individual may have a tendency to interpret social situations negatively, leading to feelings of anxiety or depression. Through CBT, they can learn to identify and challenge these negative interpretations, replacing them with more balanced and realistic thoughts.

A strength-based approach can also be integrated into CBT for neurodivergent individuals. This involves identifying and utilizing their unique strengths, interests, and abilities to enhance their sense of self-worth and empower them in their therapeutic journey.

Sensory Integration Therapy

Sensory Integration Therapy is a therapeutic approach commonly used for individuals with sensory processing difficulties, which are common among neurodivergent individuals such as those with autism or ADHD. This therapy aims to help individuals regulate and integrate sensory information more effectively.

Sensory Integration Therapy involves exposure to sensory stimuli in a controlled and structured manner with the guidance of a trained therapist. The individual engages in activities that may help them process and respond to sensory input more efficiently, ultimately improving their ability to navigate various sensory environments.

For example, a therapist may use a sensory gym equipped with different sensory experiences like swings, balance boards, or textured surfaces. The individual can engage in activities that offer a combination of calming and stimulating sensory inputs, helping them develop better sensory integration skills.

It is important to note that while Sensory Integration Therapy can be beneficial for some individuals, it may not be suitable or effective for everyone. It is crucial to tailor the therapy to the specific needs and preferences of the individual and ensure their consent and comfort throughout the process.

Mindfulness-Based Approaches

Mindfulness-based approaches have gained recognition as effective therapeutic interventions for individuals with neurodivergent conditions. Mindfulness involves paying attention to the present moment in a non-judgmental and accepting manner.

Mindfulness-based approaches, such as Mindfulness-Based Stress Reduction (MBSR) and Mindfulness-Based Cognitive Therapy (MBCT), can help neurodivergent individuals develop self-awareness, emotional regulation skills, and resilience. These approaches can be particularly helpful in managing anxiety, stress, and emotional dysregulation.

Integrating mindfulness practices into daily routines can support individuals in recognizing and managing their emotional and sensory experiences effectively. Breathing exercises, body scans, and guided meditations can cultivate a sense of centeredness and stability. Mindfulness also encourages self-compassion, promoting acceptance and understanding of one's neurodivergent traits and experiences.

It is essential to consider the individual's sensory preferences and communication style when introducing mindfulness practices. Flexibility and adaptability are key to tailoring mindfulness techniques to meet the needs and preferences of the neurodivergent individual.

Equine-Assisted Therapy

Equine-Assisted Therapy, also known as horse therapy or hippotherapy, is an experiential approach that involves interactions with horses to promote emotional, cognitive, and physical well-being. This therapy can be highly beneficial for individuals with neurodivergent conditions due to the unique qualities of horses and their non-judgmental nature.

Working with horses can help individuals develop emotional regulation, communication, and social skills. The rhythmic movements of the horse can have a calming and organizing effect, promoting relaxation and sensory integration. Interacting with horses also provides opportunities for individuals to build confidence, trust, and empathy.

Equine-Assisted Therapy typically involves activities such as grooming, feeding, and leading the horse. These activities require individuals to be present in the moment, focus on non-verbal communication, and adapt to the horse's responses. The therapist can facilitate discussions and reflections on these experiences, helping individuals develop insights into their own emotions and behaviors.

It is important for equine-assisted therapy to be conducted by trained professionals who prioritize the safety and well-being of both the individual and the horse. The therapy should be tailored to the individual's goals and preferences, ensuring a positive and empowering experience.

Resources and Caveats

When implementing therapeutic approaches for neurodivergent individuals, it is essential to consider their preferences, strengths, and unique needs. Professionals should receive specialized training to ensure the approaches are adapted appropriately and implemented in a respectful and ethical manner.

Moreover, it is crucial to foster collaboration and open communication between therapists, individuals, and their support networks. Regular feedback and evaluation can help tailor the therapeutic approaches, ensuring meaningful progress and outcomes.

Furthermore, it is important to recognize that not all therapeutic approaches will work for every neurodivergent individual. Each person is unique, and their experiences and needs may vary. It is essential to provide a range of options and resources for individuals to explore and find what works best for them.

Accessible therapeutic approaches for neurodivergent individuals play a crucial role in promoting their well-being, enhancing their abilities, and supporting their journey towards self-acceptance and fulfillment. By embracing neurodiversity and creating inclusive therapeutic environments, we can empower neurodivergent individuals and celebrate the richness of their unique perspectives and contributions.

Community Support and Advocacy

Peer Support Groups and Online Communities

In the journey of navigating neurodivergent experiences, finding a sense of community and support is incredibly valuable. Peer support groups and online communities have emerged as powerful platforms for individuals to connect with others who share their neurodivergent identities. These spaces provide a safe and inclusive environment where individuals can openly discuss their challenges, celebrate their strengths, and seek guidance from like-minded peers. In this section, we will explore the importance of peer support groups and online communities, discuss their benefits, and provide examples of popular platforms.

Benefits of Peer Support Groups

Joining a peer support group can have significant positive impacts on the mental well-being and self-acceptance of neurodivergent individuals. Here are some key benefits:

- **Validation and Understanding:** Peer support groups offer a sense of validation and understanding that comes from connecting with individuals who share similar experiences. It helps combat feelings of isolation and reassures individuals that they are not alone in their journey.

- **Mutual Support and Empowerment:** Members of peer support groups provide emotional support and practical advice to one another. Through sharing their personal experiences and strategies for coping with challenges, individuals can learn from each other and feel empowered.

- **Building Social Skills and Confidence:** Engaging with others in a supportive setting helps neurodivergent individuals develop and improve their social skills. They can practice effective communication, learn about empathy, and gain confidence in forming and maintaining relationships.
- **Education and Information Sharing:** Peer support groups serve as a valuable source of information and resources. Members can share knowledge about research, therapy options, accommodations, and advocacy efforts, leading to better-informed decisions about their own well-being.

Examples of Peer Support Groups and Online Communities

Online platforms have played a crucial role in facilitating connections between neurodivergent individuals. Here are some popular peer support groups and online communities:

- **Autistic Self-Advocacy Network Community** (ASAN): ASAN is an online community that brings together autistic individuals, providing spaces for discussion, education, and advocacy. It aims to promote acceptance and understanding of autism while emphasizing the rights of autistic individuals.
- **Wrong Planet Forum:** Wrong Planet is one of the largest online communities for individuals on the autism spectrum. It offers forums for members to discuss a wide range of topics, ask questions, and seek support.
- **ADHD Roller Coaster:** ADHD Roller Coaster is a blog and online community that focuses on ADHD-related topics. It provides support, education, and strategies for managing the challenges associated with ADHD.
- **Dyslexia Help:** Dyslexia Help is an online community that supports individuals with dyslexia and their families. It offers resources, discussion forums, and expert advice on dyslexia-related topics.
- **The Mighty:** The Mighty is a platform that includes an extensive collection of personal stories, articles, and video content related to various neurodivergent conditions. It encourages individuals to share their experiences and connect with others.

These platforms provide an outlet for neurodivergent individuals to express themselves, ask questions, seek advice, and find solidarity within a supportive community.

Challenges and Considerations

While peer support groups and online communities offer numerous benefits, it is essential to recognize and address some potential challenges:

- **Maintaining a Safe and Inclusive Space:** It is crucial to create an environment where individuals feel safe, respected, and supported. Moderators and community guidelines play a vital role in ensuring that discussions remain positive and inclusive for all members.

- **Balancing Online and Offline Interactions:** While online communities provide a convenient platform for connecting with others, it is important to strike a balance between online and offline interactions. Building relationships in person can provide additional support and foster a sense of community.

- **Avoiding Comparison and Advice Overload:** Every individual's neurodivergent journey is unique, and what works for one person may not work for another. It is important to encourage empathy, avoid comparison, and remind individuals to take advice with caution.

By recognizing and addressing these challenges, peer support groups and online communities can continue to be empowering and valuable spaces for neurodivergent individuals.

Takeaways and Further Exploration

Peer support groups and online communities offer a multitude of benefits to neurodivergent individuals, allowing them to connect with like-minded peers, share experiences, and gain valuable insights. Engaging in these spaces can enhance social skills, improve mental well-being, and foster a sense of empowerment. However, it is important to approach these spaces mindfully, ensuring inclusivity, and maintaining a healthy balance between online and offline interactions.

To further explore the world of neurodivergent communities, consider joining online platforms mentioned above and engaging in discussions. Take the opportunity to share your experiences, learn from others, and contribute to the broader neurodiversity movement. Remember, your voice and perspective matter, and by actively participating, you can make a difference in supporting and advocating for neurodiversity.

Neurodivergent Activism and Advocacy Organizations

Neurodivergent activism and advocacy organizations play a crucial role in promoting acceptance, inclusion, and equal rights for individuals with neurodivergent conditions. These organizations provide a platform for the voices and experiences of neurodivergent individuals and work towards creating a society that embraces and celebrates neurodiversity. In this section, we will explore the importance of neurodivergent activism and advocacy organizations, their key principles and goals, and some notable examples of such organizations.

Importance of Neurodivergent Activism and Advocacy Organizations

Neurodivergent activism and advocacy organizations serve as powerful agents of change in challenging the stigma, discrimination, and societal barriers faced by neurodivergent individuals. These organizations empower neurodivergent individuals to have a voice in decision-making processes that affect their lives and work towards creating a more inclusive and accepting society.

Through their efforts, these organizations raise awareness about neurodiversity and educate the public about the strengths, talents, and contributions of neurodivergent individuals. They provide support, resources, and a sense of community for neurodivergent individuals and their families, helping them navigate challenges and access necessary services.

Furthermore, neurodivergent activism and advocacy organizations actively engage in policy and legislative advocacy, striving to bring about systemic change that promotes the rights and well-being of neurodivergent individuals. They work towards creating inclusive educational environments, employment opportunities, and accessible healthcare for all.

Key Principles and Goals

Neurodivergent activism and advocacy organizations are guided by several key principles that drive their work. These principles include:

- **Self-Advocacy:** Empowering neurodivergent individuals to advocate for themselves and their rights, promoting self-determination and autonomy.

- **Acceptance and Inclusion:** Promoting a culture of acceptance and inclusion that values and respects neurodivergent individuals for who they are.

- **Equity and Social Justice:** Striving for a society that addresses systemic inequalities and ensures equal opportunities and rights for neurodivergent individuals.

- **Collaboration and Allyship:** Building partnerships and alliances with other disability rights organizations, professionals, policymakers, and the broader community to create meaningful change.

The goals of neurodivergent activism and advocacy organizations vary, but some common objectives include:

- **Raising Awareness:** Educating the public about neurodiversity, promoting understanding and dispelling misconceptions and stereotypes.

- **Policy and Legislative Advocacy:** Influencing policy and legislation to protect the rights and well-being of neurodivergent individuals and promote inclusion.

- **Support and Resources:** Providing support networks, resources, and services that meet the unique needs of neurodivergent individuals and their families.

- **Promoting Accessible Environments:** Working towards creating inclusive educational, employment, healthcare, and recreational environments that accommodate the diverse needs of neurodivergent individuals.

- **Destigmatization:** Challenging stigma and discrimination surrounding neurodivergent conditions and promoting acceptance and celebration of neurodiversity.

Notable Neurodivergent Activism and Advocacy Organizations

Several notable neurodivergent activism and advocacy organizations have emerged globally to champion the rights and well-being of neurodivergent individuals. Here are five organizations that have made significant contributions to the neurodiversity movement:

1. **Autistic Self Advocacy Network (ASAN)**

   ASAN is a prominent organization run by and for autistic individuals that focuses on promoting acceptance and social justice. ASAN advocates for policy reforms, provides resources, and facilitates leadership development programs for autistic individuals.

2. **Neurodiversity in the Workplace Initiative (NWiW)**

   NWiW is a global coalition of organizations and individuals committed to promoting the inclusion of neurodivergent individuals in the workplace. They collaborate with employers, share best practices, and advocate for inclusive hiring practices and workplace accommodations.

3. **Learning Disabilities Association of America (LDA)**

   LDA is an organization that supports individuals with learning disabilities, including dyslexia, dyscalculia, and ADHD. They provide resources, advocacy, and awareness programs to ensure that individuals with learning disabilities receive equal access to education, employment, and community participation.

4. **The National Alliance for Mental Illness (NAMI)**

   While not exclusively focused on neurodivergent individuals, NAMI is a well-known advocacy organization that provides education, support, and advocacy for individuals with mental health conditions. NAMI promotes mental health awareness and works to improve the lives of individuals with mental health conditions through policy advocacy and support programs.

5. **The Dyspraxia Foundation**

   The Dyspraxia Foundation is a UK-based organization that supports individuals with developmental coordination disorder (DCD) or dyspraxia. They provide resources, information, and raise awareness about DCD, while advocating for improved diagnosis, support, and inclusion for individuals with dyspraxia.

These organizations, along with many others, demonstrate the diversity and breadth of neurodivergent activism and advocacy efforts worldwide. They work tirelessly to promote understanding, acceptance, and equal rights for neurodivergent individuals across different domains of life.

Neurodivergent Activism and Advocacy: Going Beyond

Neurodivergent activism and advocacy can take many different forms beyond the scope of traditional organizations. Social media platforms have become powerful tools for neurodivergent individuals to connect, share experiences, disseminate knowledge, and mobilize for change. Online communities, blogs, and forums allow individuals to seek support, engage in discussions, and challenge stereotypes.

Furthermore, collaborations between neurodivergent individuals and professionals in academia, research institutions, and other domains have been instrumental in advancing the understanding of neurodiversity and developing evidence-based strategies for support and accommodation.

In conclusion, neurodivergent activism and advocacy organizations are vital for promoting neurodiversity, challenging discrimination, and creating inclusive societies. By amplifying the voices of neurodivergent individuals, these organizations drive forward positive change, foster acceptance, and empower individuals to fulfill their potential. With continued efforts and collaborations, we can move towards a future where neurodivergent individuals are fully embraced and valued for their unique strengths and contributions.

Access to Mental Health Services and Supports

Access to mental health services and supports is a critical aspect of promoting the well-being of neurodivergent individuals. Unfortunately, many barriers and challenges exist that prevent individuals from accessing the help they need. This section explores the importance of access to mental health services, the barriers faced by neurodivergent individuals, and potential solutions to enhance accessibility.

The Importance of Access to Mental Health Services

Mental health services play a crucial role in supporting the overall well-being of individuals, particularly those who are neurodivergent. These services can provide assessment, diagnosis, intervention, and ongoing support for individuals with mental health concerns. Access to such services is vital in promoting early intervention, effective treatment, and improved quality of life.

For neurodivergent individuals, access to mental health services takes added significance. Many neurodivergent individuals experience co-occurring mental health conditions, such as anxiety, depression, or ADHD, which can significantly impact their daily lives. Mental health services not only help manage these conditions but also provide a safe and supportive environment for individuals to express their feelings and concerns.

Moreover, mental health services can help individuals develop coping strategies, build resilience, and navigate social challenges. These services often offer psychoeducational programs and therapeutic interventions tailored to the specific needs of neurodivergent individuals, enhancing their overall mental well-being.

Barriers to Access

Despite the importance of mental health services, many barriers hinder access for neurodivergent individuals. These barriers can be categorized into several key areas:

1. **Financial Barriers:** The cost associated with mental health services can be a significant obstacle for many individuals, especially those with limited financial resources. Insurance coverage and affordability of services can vary, and not all services may be covered by insurance plans. This financial burden can deter individuals from seeking the help they need.

2. **Stigma and Discrimination:** Stigma surrounding mental health and neurodivergence can create significant barriers to accessing services. Many individuals fear judgment, societal prejudices, and the potential impact on their personal and professional lives. The fear of being labeled or misunderstood can prevent individuals from seeking help, leading to increased mental health challenges.

3. **Lack of Awareness and Information:** Limited knowledge and understanding of available mental health services can hinder access. Many neurodivergent individuals, their families, and even healthcare providers may not be aware of the specific services and supports tailored to their needs. This lack of information can result in missed opportunities for early intervention and support.

4. **Inadequate Mental Health Infrastructure:** Insufficient availability and accessibility of mental health services, particularly in rural or low-income areas, can pose significant challenges. Limited service providers, long waitlists, and inadequate resources can make it difficult for individuals to access timely and appropriate care.

5. **Communication and Sensory Barriers:** Communication difficulties, sensory sensitivities, and other challenges associated with neurodivergence can hinder effective communication with mental health professionals. The lack of accommodations, understanding, and accessibility adaptations can result in misunderstandings and further barriers to receiving appropriate care.

Enhancing Accessibility

Improving accessibility to mental health services for neurodivergent individuals requires a multifaceted approach. By addressing the barriers mentioned earlier, it is possible to create a more inclusive and accessible mental health system. Here are some potential solutions:

1. **Financial Support and Insurance Coverage:** Governments and healthcare systems should prioritize mental health services and ensure adequate funding to reduce financial barriers. Insurance coverage should include mental health services to make them more affordable and accessible for all individuals, regardless of their socioeconomic status.

2. **Reducing Stigma and Raising Awareness:** Efforts should be made to educate the public, healthcare professionals, and employers about neurodivergence and mental health. Anti-stigma campaigns can help combat discrimination and create a more inclusive society that encourages help-seeking behaviors.

3. **Expanding Mental Health Infrastructure:** Governments and healthcare systems should invest in expanding mental health infrastructure, particularly in underserved areas. This includes increasing the number of service providers, improving access to specialized services for neurodivergent individuals, and reducing wait times for assessments and interventions.

4. **Tailored Communication and Accommodations:** Mental health professionals should receive training on effective communication strategies and accommodations for neurodivergent individuals. This includes using plain language, visual aids, and assistive technologies to enhance communication. Sensory-friendly environments can also help reduce sensory overload and increase comfort during sessions.

5. **Collaborative Care Models:** Collaborative care approaches, involving multidisciplinary teams, can enhance accessibility and support for neurodivergent individuals. These teams may include mental health professionals, educators, social workers, and peer support specialists who work together to provide comprehensive and coordinated care.

Resources and Support

Several resources and support networks are available to help individuals overcome barriers and access mental health services. These include:

- Non-profit organizations and advocacy groups, such as the National Alliance on Mental Illness (NAMI) or Autism Self Advocacy Network, can provide information, support, and resources for neurodivergent individuals and their families.

- Online communities and forums allow individuals to connect with peers, share experiences, and seek advice on accessing mental health services.
- Community health clinics and sliding scale services offer affordable or free mental health services to individuals with limited financial resources.
- Mental health helplines and crisis centers provide immediate support and guidance for individuals in distress.
- Educational institutions, workplaces, and community organizations can offer workshops, training, and resources on mental health support and accessibility.

Conclusion

Access to mental health services and supports is crucial for promoting the well-being of neurodivergent individuals. By addressing barriers, raising awareness, and implementing inclusive practices, it is possible to enhance accessibility and ensure that all individuals can access the care they need. Creating a society that values and prioritizes mental health will contribute to a more inclusive and supportive environment for neurodivergent individuals. Remember, seeking help is not a sign of weakness but a courageous step towards self-care and personal growth.

Creating Inclusive Mental Health Policy and Practices

Creating inclusive mental health policy and practices is crucial in ensuring that individuals with neurodivergent conditions receive the support and care they need. It involves developing policies, guidelines, and strategies that promote the well-being and inclusion of neurodivergent individuals in mental healthcare systems. By addressing the unique needs and challenges faced by these individuals, we can work towards reducing stigma, improving access to services, and providing effective support for their mental health.

Understanding Neurodivergent Mental Health

Neurodivergent individuals may experience mental health conditions differently from neurotypical individuals. It is essential to recognize the specific mental health challenges they may face and the factors that contribute to their well-being. Some common mental health conditions that may co-occur with neurodivergent traits include anxiety, depression, obsessive-compulsive disorder (OCD), and attention-deficit/hyperactivity disorder (ADHD).

To create inclusive mental health policies and practices, it is vital to have a holistic understanding of how mental health impacts neurodivergent individuals. This includes recognizing the potential impact of societal stigma, healthcare disparities, and barriers to accessing appropriate support and treatment.

Reducing Stigma and Promoting Awareness

Stigma and misconceptions surrounding neurodivergent conditions can greatly impact the mental health of individuals within these communities. Inclusive mental health policy and practices should focus on reducing this stigma and promoting awareness among healthcare professionals, policymakers, and society as a whole.

Education and training programs can be developed to increase understanding and empathy towards neurodivergent individuals and their mental health. This can help address biases, challenge stereotypes, and encourage more inclusive and respectful interactions. In addition, public campaigns can be launched to raise awareness and promote acceptance of neurodiversity, emphasizing the importance of mental health support for all individuals.

Accessible and Equitable Mental Health Services

Creating inclusive mental health policy requires ensuring that neurodivergent individuals have equitable access to services and supports. This involves addressing the barriers they may face in seeking and receiving mental health care, such as financial constraints, lack of appropriate accommodations, and limited availability of specialized services.

Policymakers and service providers should work towards developing strategies that make mental health services more accessible, affordable, and culturally appropriate for neurodivergent individuals. This can include providing information and resources in accessible formats, ensuring sensory-friendly environments, and training mental health professionals to better understand and accommodate the unique needs of neurodivergent individuals.

Person-Centered Care and Support

Person-centered care is essential in inclusive mental health policy and practices. It involves tailoring mental health support to meet the individual needs, preferences, and goals of neurodivergent individuals. This approach recognizes the importance of individual autonomy and self-determination, empowering individuals to actively participate in their mental health care.

To implement person-centered care, mental health policies and practices should prioritize collaborative decision-making, respect for individual choices, and the inclusion of diverse perspectives. This can be achieved through comprehensive assessment tools, shared decision-making frameworks, and the integration of peer support and advocacy services.

Collaboration and Partnerships

Creating inclusive mental health policy and practices requires collaboration between various stakeholders, including neurodivergent individuals, healthcare professionals, policymakers, and advocacy organizations. By working together, these stakeholders can share knowledge, experiences, and expertise to shape policies and practices that better support neurodivergent mental health.

Partnerships can be established to develop and implement initiatives that promote inclusive mental health policies and practices. These partnerships can involve academic institutions, healthcare organizations, community-based agencies, and advocacy groups. Through these collaborations, resources can be pooled, and innovative approaches can be developed to address the specific needs of neurodivergent individuals.

Promoting Research and Evidence-Based Practices

Inclusive mental health policy and practices should be guided by research and evidence-based practices. This involves conducting research studies that explore the mental health needs of neurodivergent individuals, evaluate existing interventions, and develop new approaches to support their well-being.

Research can inform the development of policies and practices that are tailored to the unique strengths and challenges of neurodivergent individuals. By promoting collaboration between researchers, practitioners, and individuals with lived experience, policymakers can ensure that the mental health needs of neurodivergent individuals are accurately represented and effectively addressed.

Challenges and Considerations

Creating inclusive mental health policy and practices is not without its challenges. Some of the key considerations and challenges include:

- Funding and resource allocation: Adequate funding and resource allocation are necessary to implement inclusive mental health policies and practices. Advocacy efforts can help secure the necessary resources to support the development and implementation of these initiatives.

- Intersectionality: It is important to recognize and address the intersecting identities and experiences of individuals with neurodivergent conditions. Inclusive mental health policies and practices should consider how factors such as race, gender, sexual orientation, and socioeconomic status intersect with neurodivergent identity to shape mental health experiences and needs.

- Training and education: Healthcare professionals, policymakers, and other relevant stakeholders may require training and education to effectively implement inclusive mental health policies and practices. Training programs can help increase awareness, address biases, and provide practical strategies for supporting neurodivergent individuals' mental health.

- Evaluation and monitoring: Regular evaluation and monitoring of inclusive mental health policies and practices are essential to ensure their effectiveness and identify areas for improvement. This can involve collecting and analyzing data on service utilization, satisfaction, and outcomes among neurodivergent individuals.

In conclusion, creating inclusive mental health policy and practices is crucial for promoting the well-being and inclusion of neurodivergent individuals. By reducing stigma, ensuring equitable access to services, providing person-centered care, and promoting collaboration and research, we can work towards a mental healthcare system that effectively supports the mental health needs of neurodivergent individuals.

Promoting Neurodivergent Well-being

Promoting the well-being of neurodivergent individuals is a crucial aspect of supporting and celebrating neurodiversity. It involves creating an inclusive and supportive environment where neurodivergent individuals can thrive, achieve their full potential, and lead fulfilling lives. In this section, we will explore various strategies and approaches to promote the well-being of neurodivergent individuals in different aspects of their lives.

Understanding Neurodivergent Well-being

Neurodivergent well-being encompasses a holistic approach that addresses both the physical and mental health needs of individuals. It recognizes that neurodivergent individuals may have unique challenges and strengths that require tailored support. The promotion of well-being focuses on building resilience, self-acceptance, and fostering a positive sense of identity.

It is important to recognize that neurodivergent individuals may face higher rates of mental health conditions such as anxiety and depression. These conditions can be influenced by societal factors such as stigma, discrimination, and social isolation. Promoting neurodivergent well-being involves addressing these factors and providing appropriate support and resources.

Building Resilience and Coping Strategies

Building resilience is a key component of promoting neurodivergent well-being. Resilience refers to the ability to bounce back from challenges and adversity. Neurodivergent individuals may face unique challenges, and developing resilience can help them navigate these challenges more effectively.

Some strategies for building resilience include:

- Providing opportunities for social support and connection: Building a support network of understanding peers, mentors, and friends can provide a sense of belonging and validation.

- Encouraging self-advocacy: Empowering neurodivergent individuals to express their needs and advocate for themselves can enhance their sense of agency and self-confidence.

- Developing problem-solving skills: Teaching effective problem-solving skills can help neurodivergent individuals approach challenges with a positive mindset and find creative solutions.

- Enhancing emotional regulation skills: Providing strategies and tools for managing emotions can support neurodivergent individuals in navigating stress and anxiety.

Accessing Supportive Services

Accessing appropriate support services is essential for neurodivergent individuals to promote their well-being. These services can include mental health counseling, occupational therapy, speech therapy, and other specialized interventions tailored to individual needs.

It is important to ensure that these services are accessible and accommodating to neurodivergent individuals. This can be achieved by providing clear information about available services, removing barriers to access, and involving neurodivergent individuals and their families in the decision-making process.

Empowering Self-Advocacy

Empowering neurodivergent individuals to self-advocate is crucial for their well-being. Self-advocacy involves developing the skills and confidence to communicate their needs and preferences, assert their rights, and make informed decisions.

Some strategies for empowering self-advocacy include:

- Providing education and awareness: Educating neurodivergent individuals about their rights and promoting self-awareness can help them articulate their needs more effectively.

- Developing communication skills: Enhancing communication skills, including assertiveness and effective self-expression, can empower neurodivergent individuals to advocate for themselves.

- Promoting self-determination: Encouraging and respecting the autonomy of neurodivergent individuals enables them to participate actively in decision-making processes related to their well-being.

- Creating inclusive environments: Fostering inclusive environments that value and respect the input of neurodivergent individuals can support their confidence in advocating for their needs.

Promoting Acceptance and Inclusion

Promoting acceptance and inclusion is fundamental to supporting neurodivergent well-being. It involves challenging stereotypes, reducing stigma, and creating environments that value diversity.

Some strategies for promoting acceptance and inclusion include:

- Educating others: Raising awareness and providing education about neurodiversity can help dispel myths, reduce stigma, and foster a more inclusive society.

- Promoting inclusive policies and practices: Advocating for inclusive policies in education, employment, and healthcare settings can create a supportive environment for neurodivergent individuals.
- Encouraging empathy and understanding: Promoting empathy and understanding among peers, educators, employers, and the broader community can create a more accepting and inclusive society.
- Celebrating neurodivergent achievements: Recognizing and highlighting the strengths and contributions of neurodivergent individuals in various fields can challenge stereotypes and promote acceptance.

Improving Access to Mental Health Services

Improving access to mental health services is crucial for promoting the well-being of neurodivergent individuals. Unfortunately, many face barriers when seeking mental health support. Some strategies to improve accessibility include:

- Increasing the availability of specialized services: Providing specialized mental health services that understand and cater to the unique needs of neurodivergent individuals can ensure appropriate support.
- Supporting early intervention: Early identification and intervention can prevent mental health concerns from escalating and provide timely support for neurodivergent individuals.
- Training mental health professionals: Providing training and education for mental health professionals on neurodiversity can enhance their competence in working with neurodivergent individuals.
- Establishing peer support networks: Creating peer support networks and online communities can provide a safe space for neurodivergent individuals to connect, share experiences, and access support.

Overall, promoting the well-being of neurodivergent individuals requires a multifaceted approach that addresses individuals' unique needs, fosters resilience, empowers self-advocacy, promotes acceptance and inclusion, and improves access to appropriate services. By implementing these strategies, we can create a society that values and celebrates neurodiversity, ensuring the well-being and success of all individuals.

Neurodiversity in Media and Popular Culture

Neurodivergent Representation in Media

Portrayals of Neurodivergent Characters in Film and Television

Portrayals of neurodivergent characters in film and television have the power to shape public perception and understanding of neurodiversity. It is crucial for these portrayals to be accurate, authentic, and respectful in order to avoid perpetuating stereotypes or further stigmatizing neurodivergent individuals. In this section, we will explore the importance of accurate representation, highlight examples of positive portrayals, discuss the challenges that arise in representation, and provide strategies for improving the depiction of neurodivergent characters.

The Importance of Accurate Representation

Accurate representation of neurodivergent characters is vital for several reasons. Firstly, it helps to challenge stereotypes and misconceptions about neurodivergent individuals. By showing a diverse range of experiences and personalities, these portrayals demonstrate that neurodivergent individuals are multifaceted and not defined solely by their neurodivergent traits.

Secondly, accurate representation fosters understanding and empathy among viewers. When audiences see neurodivergent characters navigating everyday challenges, they can gain insight into the lived experiences of neurodivergent individuals. This increased understanding can lead to greater acceptance, inclusion, and support in real life.

Lastly, accurate representation provides neurodivergent individuals with role models and a sense of validation. When they see characters who share their experiences, it can help them feel seen, heard, and understood. This representation can also inspire neurodivergent individuals to pursue their passions and goals, knowing that they too can achieve success and fulfillment.

Examples of Positive Portrayals

In recent years, there have been notable examples of positive portrayals of neurodivergent characters in film and television. These examples have contributed to a more nuanced and diverse representation of neurodiversity. Here are a few notable examples:

- **"The Good Doctor"**: This TV series follows the journey of Dr. Shaun Murphy, a young surgeon with autism and savant syndrome. The show highlights Dr. Murphy's exceptional medical skills and challenges the assumptions and biases surrounding his neurodivergent identity.

- **"Atypical"**: This Netflix series revolves around the life of Sam Gardner, a high school student on the autism spectrum. The show explores Sam's relationships, aspirations, and struggles, showcasing his unique perspective and emphasizing the importance of empathy and understanding.

- **"Rain Man"**: This classic film tells the story of Charlie Babbitt, who discovers that his estranged brother Raymond, played by Dustin Hoffman, is an autistic savant. Through their journey together, the film sheds light on the strengths and challenges of neurodivergent individuals.

These examples demonstrate the power of well-rounded and authentic portrayals to challenge stereotypes and increase understanding.

Challenges in Representation

Despite the progress made in portraying neurodivergent characters, several challenges still exist. One significant challenge is the scarcity of opportunities for neurodivergent actors to play neurodivergent characters. Often, neurotypical actors are cast in these roles, which can perpetuate the exclusion and marginalization of neurodivergent individuals in the entertainment industry.

Another challenge lies in striking a balance between authenticity and entertainment value. Filmmakers and television producers sometimes face pressure to simplify or exaggerate neurodivergent traits in order to appeal to a broader audience. This can result in inaccurate and stereotypical depictions that do not accurately reflect the diversity within the neurodivergent community.

Additionally, there is a risk of fetishizing or romanticizing neurodivergent traits in media portrayals. While it is important to celebrate the strengths and unique perspectives of neurodivergent individuals, it is equally crucial to remain respectful and avoid reducing their experiences to mere plot devices or sources of inspiration for neurotypical characters.

Improving Depictions of Neurodivergent Characters

To improve the depiction of neurodivergent characters in film and television, several strategies can be employed:

- **Consultation and collaboration:** Filmmakers and television producers should actively involve neurodivergent individuals in the creative process. Consulting with neurodivergent actors, writers, and advocates can ensure that the portrayals are authentic and respectful.

- **Diverse representation:** It is important to portray the diversity of the neurodivergent community by showcasing characters with various diagnoses, backgrounds, and experiences. This helps to challenge the idea that neurodiversity is a monolithic concept and allows for more accurate and nuanced storytelling.

- **Education and awareness:** Filmmakers and television producers should invest in educating themselves and their teams about neurodiversity. This can be achieved through workshops, training sessions, and close collaboration with experts in the field. By increasing awareness, the industry can challenge biases and misconceptions.

- **Authentic storytelling:** Authenticity should be prioritized in storytelling. This involves depicting neurodivergent characters as multidimensional individuals with their own desires, dreams, and challenges, rather than reducing them to their neurodivergent traits alone. This approach allows for more nuanced and empathetic storytelling.

- **Promoting neurodivergent voices:** It is essential to amplify the voices of neurodivergent individuals in the filmmaking and television industries. This can be achieved by providing platforms for neurodivergent filmmakers, writers, and actors to share their stories and perspectives. By promoting neurodivergent voices, the industry can ensure that the narratives are authentic and representative.

By implementing these strategies, the industry can contribute to a more inclusive and accurate portrayal of neurodivergent characters in film and television.

Conclusion

Portrayals of neurodivergent characters in film and television have the potential to shape public understanding and perception of neurodiversity. Accurate and authentic representation is crucial to challenge stereotypes, foster empathy, and provide role models for neurodivergent individuals. Although progress has been made, challenges such as the underrepresentation of neurodivergent actors and the risk of simplification or fetishization of neurodivergent traits still exist. By involving neurodivergent individuals in the creative process, showcasing diverse perspectives, promoting education and awareness, prioritizing authenticity, and amplifying neurodivergent voices, the industry can contribute to a more inclusive and respectful portrayal of neurodivergent characters.

The Impact of Authentic Representation

Authentic representation in media plays a crucial role in shaping perceptions and understanding of neurodivergent individuals. It refers to the accurate portrayal of their lived experiences, challenges, strengths, and identities. When media accurately represents neurodivergent characters and their stories, it has a profound impact on society, promoting empathy, understanding, and acceptance. In this section, we will explore the importance and effects of authentic representation in various forms of media.

Promoting Empathy and Understanding

Authentic representation in media allows viewers to develop a deeper understanding of the experiences and perspectives of neurodivergent individuals. By portraying neurodivergent characters with their unique traits, interests, and challenges, media can break down stereotypes and misconceptions surrounding neurodivergence.

For example, a well-written and authentically portrayed character with autism can help demystify the condition and provide insight into the different ways individuals with autism navigate the world. This can increase empathy and understanding among viewers, helping them recognize that neurodivergent individuals have their own unique strengths and challenges.

Breaking Down Stigma and Misconceptions

Media has the power to challenge societal stigmas and misconceptions about neurodivergent individuals. By showcasing neurodivergent characters who are multi-dimensional and relatable, media can counter negative stereotypes and challenge the idea that neurodivergence is inherently a problem or a flaw.

For instance, portraying a neurodivergent character who excels in a particular field or has unique problem-solving skills can showcase the strengths and abilities of neurodivergent individuals. This can help shift public perception and challenge the notion that neurodivergence is solely a deficit or a limitation.

Providing Role Models and Representation

Authentic representation in media can provide much-needed role models for neurodivergent individuals. Seeing characters who share similar traits and experiences can offer validation and a sense of belonging, allowing neurodivergent individuals to see themselves represented positively in popular culture.

For example, a neurodivergent character who overcomes challenges and achieves success can inspire neurodivergent viewers and provide hope for their own futures. By providing diverse and authentic representations, media can empower neurodivergent individuals and help them realize their potential.

Educating and Raising Awareness

Media has the power to educate the general public about neurodivergence and raise awareness about the challenges faced by neurodivergent individuals. Through realistic portrayals, media can shed light on the neurodiversity movement, neurodivergent identities, and the importance of acceptance and inclusion.

For instance, a well-researched and accurately portrayed storyline in a television show can educate viewers about the daily struggles and triumphs of individuals with attention-deficit/hyperactivity disorder (ADHD). This can increase awareness and understanding, fostering a more inclusive and supportive society.

Challenges and Considerations

While authentic representation in media is essential, it is crucial to involve neurodivergent individuals themselves in the creation and portrayal of neurodivergent characters. Consultation with neurodivergent communities can help ensure that the representation is indeed authentic and avoids harmful stereotypes or misrepresentations.

Additionally, it is essential to recognize and address the power imbalances within the media industry. Neurodivergent voices and experiences are often underrepresented behind the scenes, limiting opportunities for authentic representation. Promoting diversity and inclusion within media production teams can help overcome these challenges and ensure that authentic representation reflects a variety of perspectives.

Conclusion

Authentic representation in media has a significant impact on promoting empathy, understanding, and acceptance of neurodivergent individuals. By accurately portraying neurodivergent characters and their stories, media can challenge stereotypes, break down stigmas, provide role models, and educate the general public. However, it is essential to involve neurodivergent individuals in the creative process and address power imbalances in the media industry to ensure that authentic representation is achieved.

In the next section, we will explore the representation of neurodiversity in literature, including autobiographies, memoirs, and fictional works that feature neurodivergent characters.

Stereotypes and Misrepresentation in Popular Culture

Popular culture, including films, television shows, books, and music, plays a significant role in shaping public perceptions and understanding of neurodivergent individuals. Unfortunately, stereotypes and misrepresentations abound in these media, perpetuating harmful misconceptions and contributing to the stigmatization of neurodiversity. In this section, we will explore the prevalence of stereotypes and misrepresentation in popular culture and discuss their impact on society.

The Power of Stereotypes

Stereotypes are oversimplified and generalized beliefs about a particular group of people. In the context of neurodiversity, stereotypes often lead to misconceptions and discrimination against neurodivergent individuals. In popular culture, these stereotypes can manifest in different ways:

1. The Eccentric Genius: One common stereotype portrays neurodivergent individuals, particularly those with autism, as brilliant savants or geniuses. While it is true that some neurodivergent individuals possess exceptional abilities in specific areas, such as mathematics or music, it is incorrect and unfair to assume

that all neurodivergent individuals possess extraordinary talents. This stereotype not only exaggerates the abilities of some individuals but also overlooks the diverse strengths and talents that exist within the neurodivergent community.

2. The Awkward Social Outcast: Another prevalent stereotype depicts neurodivergent individuals as socially inept and socially isolated. This stereotype often portrays them as unable to form meaningful relationships or engage in typical social interactions. While it is true that some neurodivergent individuals may face challenges in social situations, it is important to recognize that social abilities and preferences can vary widely within the neurodivergent community. By perpetuating this stereotype, popular culture further isolates neurodivergent individuals and reinforces societal biases.

3. The Violent or Dangerous Individual: Perhaps one of the most damaging stereotypes is the portrayal of neurodivergent individuals as dangerous or prone to violence. This stereotype is often sensationalized in crime dramas and horror movies, perpetuating fear and misunderstanding. It is crucial to debunk this stereotype and emphasize that neurodivergent individuals are no more likely to engage in violent behavior than the general population. By portraying them as dangerous, popular culture contributes to the marginalization and discrimination they face in their everyday lives.

Implications and Consequences

The misrepresentation of neurodivergent individuals in popular culture has severe consequences for both individuals and society as a whole. These include:

1. Reinforcing stereotypes and stigma: When popular culture perpetuates stereotypes about neurodivergent individuals, it reinforces existing stigmas and biases, making it more challenging for them to be accepted and included in society. These stereotypes contribute to the social and economic marginalization of neurodivergent individuals, limiting their opportunities for education, employment, and relationships.

2. Alienation and isolation: Misrepresentations in popular culture can lead to feelings of alienation and isolation among neurodivergent individuals. Seeing themselves portrayed negatively or inaccurately in media can erode their self-esteem and sense of belonging. This can have a significant impact on their mental health and overall well-being.

3. Lack of understanding and empathy: Stereotypes and misrepresentations perpetuated in popular culture contribute to a lack of understanding and empathy towards neurodivergent individuals. This lack of understanding hinders efforts to

create inclusive and supportive environments in schools, workplaces, and communities.

Challenging Stereotypes

To address the issue of stereotypes and misrepresentation in popular culture, it is essential to challenge and counteract these portrayals. Here are some strategies that can help promote accurate and positive representations of neurodivergent individuals:

1. Authentic representation: Popular culture should prioritize authentic representation of neurodivergent individuals by involving them in the creation and production process. This can help ensure that their experiences and perspectives are accurately portrayed.

2. Diverse narratives: Popular culture should showcase the diversity within the neurodivergent community by telling a wide range of stories. It is crucial to move beyond the narrow focus on portraying neurodivergent individuals as either geniuses or socially inept, and instead, explore the complexities and nuances of their lives.

3. Collaboration with advocacy organizations: Media creators and producers should collaborate with neurodivergent advocacy organizations and individuals to gain insights and consultation on accurate portrayals. This collaborative approach can lead to more authentic and respectful representations.

4. Education and awareness: It is essential to increase public awareness and understanding of neurodiversity to challenge existing stereotypes. Media platforms can play a significant role in this regard by incorporating accurate information about neurodivergent conditions and experiences into their content.

5. Celebrating neurodivergent achievements: Popular culture should highlight the achievements and contributions of neurodivergent individuals to counteract negative stereotypes. By celebrating their talents and successes, we can challenge the notion that neurodivergence is solely characterized by deficits.

In conclusion, stereotypes and misrepresentation in popular culture have significant implications for neurodivergent individuals. By promoting accurate and positive portrayals, we can challenge these stereotypes, increase understanding and empathy, and create a more inclusive society that celebrates the diversity of neurodiversity. It is the responsibility of media creators, consumers, and society as a whole to actively challenge and rectify the misrepresentation of neurodiversity in popular culture.

Neurodivergent Artists and Creatives

Neurodivergent individuals have a unique perspective and experience of the world, which often translates into exceptional artistic and creative abilities. In this section, we explore the contributions of neurodivergent artists and creatives across various artistic mediums, highlighting their creative processes, artistic styles, and the impact of their work.

The Power of Neurodivergent Art

Neurodivergent art challenges traditional norms and conventions, offering fresh and alternative perspectives that reflect the rich diversity of human experience. It often explores themes related to neurodiversity, personal identity, social justice, inclusion, and the complexity of human emotions. Through their art, neurodivergent artists aim to express themselves authentically, raise awareness about neurodivergent experiences, and promote acceptance and understanding.

Neurodivergent art has the power to evoke strong emotions, provoke thought, and challenge societal perceptions. It provides a platform for neurodivergent individuals to communicate their experiences and advocate for change. By sharing their unique perspectives, neurodivergent artists contribute to a more inclusive and diverse artistic landscape.

Autistic Art and Creativity

Autistic artists are known for their exceptional attention to detail, their ability to notice patterns, and their unique ways of perceiving the world. Many autistic artists have a strong affinity for visual art, showcasing their talents in painting, drawing, photography, and digital art.

The autistic art community celebrates various art styles, including realistic, abstract, and surrealistic representations. Some artists possess a remarkable ability to portray intricate details and intricate patterns, while others explore their imagination through colorful and vivid abstract art.

One prominent autistic artist is Stephen Wiltshire, who is known for his detailed, panoramic cityscape drawings. Wiltshire has an extraordinary photographic memory that allows him to accurately recreate complex cityscapes entirely from memory. His artwork demonstrates the power of neurodivergent minds and their unique ways of perceiving and representing the world.

Creativity and Asperger's Syndrome

Individuals with Asperger's Syndrome often possess a strong passion for their artistic pursuits and display remarkable creative abilities. Many individuals with Asperger's excel in fields such as music, writing, poetry, and computer programming.

Musical talent is prevalent among individuals with Asperger's, with many excelling in playing instruments or composing music. Their attention to fine details, ability to concentrate for extended periods, and heightened sensitivity to sound often contribute to their musical abilities. Hans Zimmer, the renowned film composer, openly discusses his experiences as an individual with Asperger's and the role his condition plays in his creative process.

In addition to music, individuals with Asperger's often demonstrate a talent for writing and poetry. Their unique perspectives, attention to detail, and extensive knowledge in specific areas of interest contribute to their distinctive literary style. Temple Grandin, a prominent autism advocate, author, and professor, showcases the power of Asperger's creativity through her insightful writings on animal behavior and her personal experiences.

Diverse Perspectives in the Art World

Neurodivergent artists bring diverse perspectives and experiences to the art world, challenging traditional notions of aesthetics and artistic expression. Their work often pushes boundaries and offers new interpretations of art forms.

One such artist is Chuck Close, who is known for his large-scale, photorealist portraits. Close is living with prosopagnosia, a condition that impairs his ability to recognize faces. His artwork focuses on portraying faces with intricate details, highlighting the complexity of human identity beyond facial recognition. Close's work challenges viewers to reconsider the way they perceive and define others.

Another example is Judy Singer, an autistic artist, and author who coined the term "neurodiversity" in the late 1990s. Singer's artwork often explores neurodivergent experiences and challenges societal stereotypes. Through her creative expression, Singer advocates for acceptance and celebrates the strengths and contributions of neurodivergent individuals.

Art Therapy and Neurodivergent Individuals

Art therapy provides a powerful outlet for neurodivergent individuals to explore their emotions, improve self-expression, and enhance their overall well-being. Through various art forms, individuals can communicate their thoughts, feelings, and experiences in a non-verbal and often more accessible way.

Art therapy sessions can be tailored to meet the specific needs of neurodivergent individuals, providing them with a safe space to express themselves freely. Therapists trained in working with neurodivergent individuals use art materials and creative techniques to encourage self-reflection, emotional awareness, and personal growth.

Art therapy has been shown to improve self-esteem, reduce anxiety, enhance emotional regulation, and promote self-discovery among neurodivergent individuals. By embracing their creative potential, neurodivergent individuals can engage in a form of therapy that is both enjoyable and beneficial for their mental health and well-being.

Conclusion

Neurodivergent artists and creatives bring a unique and valuable perspective to the art world. Their work challenges traditional norms, promotes acceptance and understanding, and celebrates the diversity of human experience. Through art, neurodivergent individuals communicate their thoughts, emotions, and personal narratives, contributing to a more inclusive and diverse artistic landscape. Art therapy provides a powerful tool for neurodivergent individuals to explore their emotions, enhance self-expression, and improve overall well-being. Moving forward, it is essential to continue supporting and amplifying the voices of neurodivergent artists and creatives, ensuring their contributions are recognized and celebrated.

Media as a Platform for Neurodiversity Advocacy

Media plays a significant role in shaping public perception and understanding of neurodiversity. It has the power to challenge stereotypes, promote awareness, and advocate for the rights and inclusion of neurodivergent individuals. This section explores how media can serve as a platform for neurodiversity advocacy and highlights examples of successful initiatives in this area.

Challenging Stereotypes and Promoting Awareness

One of the key roles of media in neurodiversity advocacy is to challenge stereotypes and promote awareness. By featuring authentic and diverse portrayals of neurodivergent individuals, media can challenge misconceptions and foster a more accurate understanding of their experiences.

Documentaries, films, and television shows have the potential to provide powerful insights into the lives of neurodivergent individuals. For example, the documentary "Life, Animated" explores the story of Owen Suskind, a young man with autism who uses Disney movies to communicate and navigate the world. This documentary not only showcases the unique strengths and challenges of autism but also emphasizes the importance of embracing individuality and supporting neurodivergent individuals in their journey.

In addition to documentaries, fictional works can also contribute to neurodiversity advocacy. Books such as "The Curious Incident of the Dog in the Night-Time" by Mark Haddon and "Marcelo in the Real World" by Francisco X. Stork have gained widespread popularity for their insightful and empathetic portrayals of neurodivergent characters. These narratives allow readers to develop a deeper understanding of the experiences and perspectives of neurodivergent individuals, encouraging empathy and dismantling stereotypes.

Amplifying Neurodivergent Voices

Media platforms can be used to amplify the voices of neurodivergent individuals themselves. Through blogs, vlogs, podcasts, and social media, neurodivergent individuals can share their experiences, advocate for their rights, and contribute to the broader neurodiversity movement.

One inspiring example is the blog "Thinking Person's Guide to Autism," co-founded by Shannon Des Roches Rosa. This platform provides a space for autistic individuals, parents, and professionals to share their stories, insights, and expertise. By amplifying neurodivergent voices, the blog challenges dominant narratives and fosters a more nuanced understanding of autism.

Online communities and social media hashtags, such as #ActuallyAutistic, provide forums for neurodivergent individuals to connect, support one another, and advocate for change. These platforms enable neurodivergent individuals to unite, share resources, and raise awareness about their unique strengths, challenges, and needs.

Advocacy Through Popular Culture

Popular culture, including music, art, and performance, can also serve as a medium for neurodiversity advocacy. Neurodivergent artists have been able to use their creativity and unique perspectives to challenge societal norms and perceptions.

Music, in particular, has been harnessed as a therapeutic and expressive outlet for neurodivergent individuals. Artists like Susan Boyle, who gained fame through her appearance on "Britain's Got Talent," have demonstrated the immense talent and creativity that neurodivergent individuals can bring to the world of music. Their success highlights the importance of recognizing and celebrating diverse forms of talent and expression.

Visual and performing arts have also provided a platform for neurodivergent artists to share their unique perspectives and challenge societal norms. The work of artists such as Temple Grandin, an autistic professor and advocate, has not only revolutionized our understanding of animal behavior but also inspired neurodivergent individuals to pursue their passions and contribute to the arts.

Promoting Inclusivity and Advocacy Skills

Media can also play a role in promoting inclusivity and equipping neurodivergent individuals with advocacy skills. Through storytelling and representation, media can inspire neurodivergent individuals to embrace their identities, develop self-advocacy skills, and fight for their rights.

TV shows like "Atypical" and "The Good Doctor" depict neurodivergent characters navigating the challenges of life, relationships, and work. These shows not only raise awareness but also empower neurodivergent individuals by providing relatable role models and highlighting the importance of self-advocacy.

Additionally, media can provide resources and information for neurodivergent individuals and their allies. Websites, podcasts, and YouTube channels dedicated to neurodiversity advocacy offer valuable insights, tips, and resources for navigating various aspects of life, including education, employment, and mental health.

Conclusion

Media is a powerful tool for neurodiversity advocacy, providing a platform to challenge stereotypes, promote awareness, amplify neurodivergent voices, and empower individuals to advocate for their rights. By embracing diverse narratives and perspectives, media can help create a more inclusive and accepting society. It is essential that we continue to support and encourage media initiatives that celebrate neurodiversity and promote the inclusion and well-being of all individuals, regardless of their neurological differences.

Neurodiversity in Literature

Autobiographies and Memoirs of Neurodivergent Individuals

Autobiographies and memoirs written by neurodivergent individuals provide personal accounts of their experiences, challenges, strengths, and perspectives.

These narratives are invaluable in promoting understanding, empathy, and acceptance of neurodiversity. They offer a glimpse into the rich inner worlds of individuals who may think, learn, and perceive differently from the neurotypical population. By sharing their stories, neurodivergent individuals become advocates for their unique identities, paving the way for greater inclusion and appreciation in society.

The Power of Personal Stories

Autobiographies and memoirs have a profound impact on readers as they offer an intimate and authentic portrayal of an individual's experiences. They allow us to gain insight into the day-to-day struggles, triumphs, and emotions of neurodivergent individuals. By hearing these stories in their own words, we can challenge our preconceived notions and develop a deeper understanding of their lived realities.

One important aspect of these personal stories is that they defy stereotypes and misconceptions about neurodivergent individuals. They provide a counter-narrative to the prevailing medical model approach that pathologizes and marginalizes neurodivergent traits. As readers engage with these narratives, they witness the vast diversity of experiences within the neurodivergent community and recognize that neurodivergent individuals have unique strengths and perspectives to offer.

Neurodivergent Autobiographies: Shaping Perspectives

Neurodivergent autobiographies and memoirs cover a range of neurodevelopmental conditions, including autism, ADHD, dyslexia, and intellectual disabilities. These narratives offer personal insights into the challenges and triumphs associated with each condition, providing a nuanced and individualized perspective that challenges broad generalizations.

For example, temple Grandin's memoir "Thinking in Pictures" gives readers a first-hand account of what it is like to experience the world as an individual with autism. She eloquently describes her sensory experiences, her struggles with social communication, and the unique way she processes information. Through her story, readers gain a deeper appreciation for the strengths and talents that can accompany autism, such as her exceptional visual thinking abilities and her groundbreaking contributions to animal welfare.

Similarly, John Elder Robison's memoir "Look Me in the Eye" provides an insight into the life of an individual living with Asperger's syndrome. Robison shares his experiences growing up, navigating relationships, and finding his place in the world. His story showcases the value of accommodation and acceptance in enabling neurodivergent individuals to reach their full potential.

Memoirs with Intersectionality

Neurodivergent individuals also exist within various intersecting identities that shape their experiences; autobiographies and memoirs exploring these intersections provide important perspectives. For instance, "Pretending to be Normal" by Liane Holliday Willey discusses her experiences as an autistic woman. She addresses the unique challenges faced by women on the autism spectrum, such as the tendency for their symptoms to be overlooked or misdiagnosed due to gender biases in diagnostic criteria.

Another example is "Black Girl, White School: Thriving, Surviving, and No, You Can't Touch My Hair" by Tiffany Jackson. This memoir explores the intersection of neurodivergence and race, shedding light on the experiences of a Black neurodivergent individual and the additional barriers and discrimination she faces. This intersectional perspective challenges the notion that neurodivergent experiences are uniform and emphasizes the importance of acknowledging and addressing the multiple dimensions of individuals' identities.

Impact and Advocacy

Neurodivergent autobiographies and memoirs have helped pave the way for greater acceptance and inclusion in society. These personal stories play a crucial role in debunking myths and misconceptions surrounding neurodivergent individuals. By sharing their experiences, neurodivergent authors become advocates, challenging societal norms, and advocating for change.

Additionally, these narratives can be powerful tools for self-reflection and growth for neurodivergent individuals themselves. By seeing their experiences reflected in the stories of others, they can gain a sense of validation, build self-esteem, and find inspiration in navigating their own paths.

Resources and Recommendations

Here are a few recommended autobiographies and memoirs written by neurodivergent authors:

- "NeuroTribes: The Legacy of Autism and the Future of Neurodiversity" by Steve Silberman.

- "Born on a Blue Day: A Memoir of Asperger's and an Extraordinary Mind" by Daniel Tammet.
- "Uniquely Human: A Different Way of Seeing Autism" by Barry M. Prizant.
- "The Reason I Jump: The Inner Voice of a Thirteen-Year-Old Boy with Autism" by Naoki Higashida.
- "Nobody Nowhere: The Extraordinary Autobiography of an Autistic" by Donna Williams.

These memoirs offer diverse perspectives and provide valuable insight into the unique experiences of neurodivergent individuals. They serve as resources for individuals seeking to understand neurodiversity, educators supporting neurodivergent students, and professionals working in various fields related to neurodiversity.

Conclusion

Autobiographies and memoirs of neurodivergent individuals are a vital resource for promoting understanding and acceptance of neurodiversity. By sharing their experiences through personal narratives, neurodivergent authors challenge stereotypes, provide insights into their unique perspectives, and advocate for greater inclusion in society. These stories create opportunities for readers to develop empathy, question their assumptions, and embrace the diversity of human cognition and experience. As the field of neurodiversity continues to evolve, these personal stories will play a vital role in shaping policies, practices, and attitudes towards neurodivergent individuals.

Fictional Works that Feature Neurodivergent Characters

In the realm of literature and storytelling, fictional works have the power to shed light on neurodivergent experiences and challenge societal perceptions. Through empathetic and authentic portrayals, these works contribute to the understanding and acceptance of neurodiversity. In this section, we will explore some notable fictional works that feature neurodivergent characters.

1. "The Curious Incident of the Dog in the Night-Time" by Mark Haddon

"The Curious Incident of the Dog in the Night-Time" is a captivating novel written by Mark Haddon. The story is narrated by Christopher Boone, a 15-year-old boy with autism spectrum disorder (ASD). Christopher has a remarkable talent for mathematics and an extraordinary ability to observe details, yet struggles with social interactions and sensory overload.

The novel takes the reader on a journey inside Christopher's mind as he investigates the mysterious death of a neighbor's dog. Haddon skillfully portrays Christopher's unique perspective, providing insights into the sensory sensitivities, rigid thought patterns, and intense emotional experiences that individuals with ASD often face. Through this engaging narrative, readers gain a deeper understanding of the challenges and strengths associated with autism.

2. "The Rosie Project" by Graeme Simsion

"The Rosie Project" by Graeme Simsion introduces us to Don Tillman, a genetics professor who exhibits traits consistent with Asperger's syndrome (now classified under the broader term of ASD). Don is highly intelligent, logical, and regimented in his daily routines. However, his social awkwardness and difficulty understanding social cues make it challenging for him to form romantic relationships.

The novel follows Don's humorous and heartfelt journey as he embarks on "The Wife Project" – a scientific approach to finding a suitable life partner. As readers witness Don's interactions with various characters, they gain insight into the complexities of living with neurodivergent traits while appreciating the unique perspectives and strengths that individuals like Don bring to the world.

3. "Eleanor Oliphant Is Completely Fine" by Gail Honeyman

In Gail Honeyman's "Eleanor Oliphant Is Completely Fine," the eponymous character Eleanor Oliphant exhibits several characteristics associated with neurodivergence, including social difficulties, adherence to routines, and intense focus on specific interests. Eleanor is a loner who follows a strict schedule and struggles with understanding and participating in social norms.

As the story unfolds, readers delve into Eleanor's past trauma and witness her personal growth and transformation. Honeyman's portrayal of Eleanor's neurodivergent experiences is both insightful and compassionate, highlighting the impact of trauma on mental health and emphasizing the importance of community and authentic connections in fostering well-being.

4. "The Speed of Dark" by Elizabeth Moon

"The Speed of Dark" by Elizabeth Moon is a thought-provoking science fiction novel set in the near future. The story revolves around Lou Arrendale, an autistic man who is presented with the opportunity to undergo an experimental treatment that could potentially "cure" his autism. As Lou contemplates this decision, he grapples with questions related to identity, ethics, and the acceptance of neurodivergent individuals in society.

The novel delves into Lou's experiences and inner thoughts, offering readers an intimate glimpse into the challenges and strengths associated with autism. Moon's work prompts us to reflect on the importance of embracing neurodiversity and challenging the notion that there is a "normal" way of being.

5. "The Kiss Quotient" by Helen Hoang

"The Kiss Quotient" by Helen Hoang is a compelling romance novel that features Stella Lane, a brilliant econometrician with Asperger's syndrome. Stella is highly focused on her career but struggles with intimacy and relationships. To gain more experience in the realm of dating, she hires an escort named Michael Phan.

Hoang's novel explores Stella's desires, vulnerabilities, and personal growth as she navigates the complexities of love. Through Stella's perspective, readers gain insight into the sensory sensitivities, social difficulties, and atypical thought processes often associated with autism. "The Kiss Quotient" challenges stereotypes and highlights the diverse romantic experiences of neurodivergent individuals.

Conclusion

Fictional works that feature neurodivergent characters provide valuable insights into the lived experiences of individuals with conditions such as autism spectrum disorder and Asperger's syndrome. By immersing readers in the perspectives of these characters, these stories foster empathy, challenge stereotypes, and promote understanding and acceptance of neurodiversity.

Through the power of storytelling, authors like Mark Haddon, Graeme Simsion, Gail Honeyman, Elizabeth Moon, and Helen Hoang create narratives that amplify neurodivergent voices and contribute to the diverse representation of neurodiversity in literature. As these stories find their way into the hands of readers, they serve as catalysts for meaningful conversations and promote a more inclusive and accepting society.

Literary Critique and the Neurodiversity Movement

Literary critique plays a crucial role in shaping our understanding of the neurodiversity movement and its impact on society. Through the analysis of literary works featuring neurodivergent characters and themes, we can gain insights into the challenges, experiences, and perspectives of individuals with diverse neurological conditions. This section explores how literary critique contributes to the discourse surrounding neurodiversity and highlights the significance of storytelling in promoting empathy, understanding, and acceptance.

Literary Representations of Neurodiversity

Literature provides a platform for diverse neurodivergent voices to be heard and acknowledged. Authors have used their creative talents to depict characters with various neurodevelopmental conditions, such as autism spectrum disorder, ADHD, dyslexia, and intellectual disabilities, among others. These literary representations aim to illuminate the rich complexity of neurodivergent experiences and challenge the stereotypes and misconceptions often associated with these conditions.

One example of a literary work that explores neurodiversity is "The Curious Incident of the Dog in the Night-Time" by Mark Haddon. The novel presents the narrative from the perspective of Christopher Boone, a fifteen-year-old boy with Asperger's syndrome. By delving into Christopher's thoughts and emotions, Haddon provides readers with an intimate understanding of the challenges and strengths associated with autism spectrum disorder.

Another notable work is "Marcelo in the Real World" by Francisco X. Stork, which tells the story of Marcelo Sandoval, a seventeen-year-old with an autism-like condition. The novel delves into the intricacies of Marcelo's world, exploring his unique way of thinking and his struggle to navigate the complexities of the real world. Through Marcelo's journey, Stork sheds light on the importance of acceptance and finding one's place in society.

Critiquing Neurodiversity Representations

Literary critique provides a platform for evaluating and discussing the portrayals of neurodivergent characters and themes. It allows readers, academics, and scholars to analyze the accuracy, authenticity, and ethical dimensions of these representations. By critically examining the narratives, language, and character development, we can identify both strengths and limitations in the depiction of neurodiversity in literature.

One aspect of critique involves interrogating the extent to which neurodivergent characters are portrayed as fully-rounded individuals, rather than one-dimensional stereotypes. It is essential to assess whether the representation aligns with the lived experiences and perspectives of individuals with neurodivergent conditions. Does the portrayal capture the internal struggles, emotions, and nuances associated with these conditions, or does it rely on clichés and simplifications?

Furthermore, literary critique helps us assess whether neurodivergent characters are given agency and autonomy in the narrative. Are their perspectives valued and respected, or are they relegated to passive or tokenistic roles? Examining power dynamics within the storytelling framework allows us to uncover any biases, ableism, or marginalization that may exist.

The Power of Neurodivergent Storytelling

Neurodivergent storytelling plays a crucial role in advocating for the inclusion and acceptance of individuals with diverse neurological conditions. Literature allows readers to step into the shoes of neurodivergent characters, fostering empathy and understanding. Through the power of storytelling, these narratives challenge societal norms and challenge mainstream perceptions of neurodiversity.

By exposing readers to diverse neurodivergent experiences, literature has the potential to challenge ableist beliefs and dismantle stereotypes. It encourages readers to question preconceived notions and recognize the value and contributions of neurodivergent individuals in society.

Moreover, neurodivergent storytelling can provide a sense of validation and empowerment for individuals with neurodevelopmental conditions. Seeing themselves represented positively in literature can provide a sense of belonging, foster self-acceptance, and inspire others to embrace their own neurodivergent identities.

Challenges and Future Directions

While literary critique and neurodivergent storytelling have made significant strides in promoting the neurodiversity movement, challenges remain. The diversity within the neurodivergent community demands a multiplicity of narratives, encompassing a broad range of experiences, cultures, and intersecting identities. It is crucial to ensure that neurodivergent voices from marginalized communities are amplified and represented authentically.

Additionally, it is important to consider the ethical implications of storytelling, especially when neurodivergent characters are created by neurotypical authors. Sensitivity, respect, and accuracy are paramount in portraying neurodivergent experiences, and authors should engage in extensive research and consultation with neurodivergent individuals to avoid perpetuating harmful stereotypes or misrepresentations.

Overall, the intersection of literary critique and the neurodiversity movement opens up new possibilities for fostering understanding, empathy, and acceptance in society. By engaging with diverse narratives and critically analyzing their portrayal, we can promote a more inclusive and equitable world for individuals of all neurological backgrounds.

Further Reading

1. Silberman, S. (2015). Neurotribes: The Legacy of Autism and the Future of Neurodiversity. New York: Penguin Random House.

2. Milton, D. (Ed.). (2012). Autism and Representation. London: Routledge.

3. Casares, R. (2011). Asperger's on the Inside: A Memoir. London: Jessica Kingsley Publishers.

Neurodiversity and Intersectionality in Literature

Neurodiversity and intersectionality are two important concepts that have gained significant attention in recent years. In the context of literature, these concepts play a crucial role in shaping our understanding of diverse identities and experiences. This section explores the intersectionality of neurodiversity in literature, highlighting the ways in which authors incorporate neurodivergent characters and explore their multiple dimensions of identity.

Understanding Intersectionality

Intersectionality is a framework that recognizes how different aspects of a person's identity, such as race, gender, sexuality, and disability, intersect and influence their experiences. It acknowledges that individuals can hold multiple marginalized identities, and these intersecting identities shape their social, cultural, and political realities.

In the context of neurodiversity, intersectionality recognizes that neurodivergent individuals may also belong to other marginalized groups, such as racial and ethnic minorities, LGBTQ+ communities, or individuals with disabilities. This intersectionality can influence their experiences of neurodivergence and shape how they navigate societal structures and systems.

Representation of Neurodiversity in Literature

Literature has the power to shape narratives and challenge societal perceptions. Authors often include neurodivergent characters in their works to provide diverse perspectives and promote empathy and understanding. These neurodivergent characters add depth and complexity to the stories they inhabit, allowing readers to explore the intersectionality of their identities.

Neurodivergent characters in literature can range from protagonists to secondary characters, and their experiences may vary widely. Some authors choose to highlight the challenges faced by neurodivergent individuals, while others focus on their strengths and resilience. Regardless of the approach, the inclusion of neurodivergent characters in literature contributes to the broader neurodiversity movement by challenging stereotypes and promoting acceptance.

Examining Intersectionality in Literature

When exploring intersectionality in literature, it is essential to consider how the intersection of neurodivergence with other marginalized identities influences the characters' experiences. This examination allows readers to gain a deeper understanding of the complex ways in which intersections shape neurodivergent individuals' lives.

For example, a novel may feature a Black autistic character who faces not only the challenges of being neurodivergent but also the realities of systemic racism. This intersectionality highlights the layered discrimination and unique experiences that result from the convergence of multiple marginalized identities. By exploring these intersections, literature can challenge dominant narratives and provide a platform for underrepresented voices.

Neurodivergent Authors and Own Voices Narratives

The concept of "own voices" has emerged within literature, emphasizing the importance of authors from marginalized communities telling their own stories. In the context of neurodiversity, own voices narratives are those written by neurodivergent authors who draw from their lived experiences.

Neurodivergent authors bring authentic and nuanced perspectives to their work, providing readers with a deeper understanding of neurodivergent experiences. These narratives challenge stereotypes and shed light on the diverse realities of neurodivergent individuals. By supporting and amplifying neurodivergent voices in literature, we can foster greater empathy, understanding, and acceptance.

Promoting Intersectionality and Neurodiversity in Literature

Promoting intersectionality and neurodiversity in literature involves several key considerations. Authors must be mindful of representing the diverse experiences of neurodivergent individuals, elevating marginalized voices, and avoiding harmful stereotypes. Sensitivity readers from neurodivergent backgrounds can provide valuable insights and ensure authentic representation.

Educators and readers can actively seek out literature that embraces intersectionality and neurodiversity. By diversifying our reading lists and supporting neurodivergent authors, we contribute to a more inclusive literary landscape. Engaging in critical discussions about the intersectionality portrayed in literature can also expand our understanding of diverse life experiences and foster empathy.

Conclusion

The intersectionality of neurodiversity in literature plays a vital role in representing the diverse lived experiences of neurodivergent individuals. By incorporating neurodivergent characters and exploring the convergence of multiple marginalized identities, literature challenges stereotypes and promotes empathy and understanding. Through own voices narratives and the promotion of intersectional perspectives, literature can contribute to a more inclusive and accepting society.

The Power of Storytelling in Shaping Perceptions of Neurodivergent Individuals

Storytelling is a powerful tool that can shape the way people perceive and understand the world around them. In the context of neurodiversity, storytelling plays a crucial role in challenging stereotypes, promoting acceptance, and fostering empathy towards neurodivergent individuals. Through narratives and personal accounts, stories have the ability to humanize the experiences of those who are neurodivergent and create a bridge of understanding between neurotypical and neurodivergent communities.

Challenging Stereotypes and Misconceptions

One of the main functions of storytelling in the realm of neurodiversity is to challenge stereotypes and misconceptions about neurodivergent individuals. Traditional narratives often present neurodivergent characters as one-dimensional, emphasizing their deficits rather than their strengths. This perpetuates the stigma and alienation faced by neurodivergent individuals in society.

By sharing diverse and authentic stories, we can debunk these misconceptions and provide a more accurate representation of the neurodivergent experience. Through storytelling, we can showcase the unique talents, perspectives, and contributions of neurodivergent individuals, highlighting their strengths rather than focusing solely on their challenges.

Fostering Empathy and Understanding

Storytelling has the power to evoke empathy and foster understanding among its audience. When people are exposed to narratives that explore the inner world and lived experiences of neurodivergent individuals, they may develop a deeper understanding of the challenges they face and the unique way they perceive the world.

Through storytelling, neurotypical individuals can gain insight into the sensory sensitivities, social challenges, and cognitive differences that neurodivergent individuals may experience. This understanding can cultivate empathy, compassion, and acceptance towards neurodivergent individuals, breaking down barriers and promoting inclusivity.

Providing a Platform for Neurodivergent Voices

Storytelling also provides a platform for neurodivergent individuals to share their own stories, perspectives, and experiences. By amplifying neurodivergent voices, we can empower them to become agents of change and advocates for neurodiversity.

Through autobiographies, memoirs, and other forms of literature, neurodivergent individuals can share their own narratives, challenging societal norms and fostering a sense of agency. This not only allows neurodivergent individuals to take control of their own narratives but also provides an opportunity for the wider society to listen and learn from their lived experiences.

Creating Connection and Building Networks

Storytelling can also create connections and build networks within the neurodivergent community and beyond. When neurodivergent individuals see themselves reflected in stories, they may feel a sense of validation and belonging. This can help combat feelings of isolation and promote self-acceptance.

Furthermore, storytelling can encourage the formation of support networks and communities for neurodivergent individuals. By sharing their narratives and experiences, individuals with similar backgrounds can connect, find common ground, and offer support and understanding to one another.

Changing the Narrative

The power of storytelling lies not only in shaping perceptions but also in changing the narrative around neurodivergent individuals. By actively promoting and seeking out diverse stories, we can challenge the dominant discourse and create a more inclusive society.

By incorporating neurodivergent characters and narratives into mainstream media, literature, and other forms of popular culture, we can normalize neurodiversity and create a world where neurodivergent individuals are seen as valuable contributors to society.

Unconventional Approaches: Neurodivergent Storytelling Techniques

In addition to traditional storytelling formats, there is a growing movement of neurodivergent individuals exploring unconventional storytelling techniques that celebrate their neurodivergent identity. For example, some individuals with autism may express themselves through visual storytelling, using visual symbols and graphics to communicate their experiences and emotions.

These unconventional approaches not only provide a means of self-expression for neurodivergent individuals but also offer unique perspectives and insights that challenge traditional storytelling conventions.

Resources and Practices

There are several resources and practices that can support the power of storytelling in shaping perceptions of neurodivergent individuals. One such resource is the Neurodivergent Storytellers project, which aims to amplify the voices of neurodivergent individuals through various mediums, including books, podcasts, and online platforms.

Additionally, inclusive storytelling techniques, such as person-first language and avoiding ableist language, can further promote respect and dignity for neurodivergent individuals.

Conclusion

Storytelling is a compelling and transformative tool in the fight for neurodiversity acceptance and inclusion. By challenging stereotypes, fostering empathy, amplifying neurodivergent voices, creating connections, changing narratives, and exploring unconventional approaches, storytelling has the power to shape perceptions, break down barriers, and build a more just and inclusive society.

As we continue to embrace the power of storytelling, let us celebrate the rich diversity of neurodivergent experiences and work towards a world where everyone's story is valued and heard.

Neurodivergent Voices in Music, Art, and Performance

Music as a Therapeutic and Expressive Outlet

Music has long been recognized as a powerful tool for healing and self-expression. It has the ability to transcend language and cultural barriers, reaching deep within the human soul. In the context of neurodiversity, music can play a particularly significant role in supporting and empowering individuals with neurodevelopmental conditions. This section will explore the therapeutic and expressive benefits of music for neurodivergent individuals and provide insights into how music can be effectively used as a tool for emotional regulation, communication, and personal growth.

Music Therapy: Enhancing Well-being and Communication

Music therapy is a specialized field that harnesses the therapeutic properties of music to improve the well-being of individuals with various neurological and psychological conditions. It is an evidence-based practice that utilizes both active and passive engagement with music to achieve therapeutic goals. In the context of neurodiversity, music therapy can provide a safe and non-threatening environment for individuals to express themselves, develop social skills, and enhance emotional regulation.

Musical activities in therapy sessions may include singing, playing musical instruments, movement to music, songwriting, and listening to music. These activities can support individuals with neurodivergent traits in various ways:

- **Emotional Regulation:** Music has the power to evoke and regulate emotions. It can help individuals with neurodivergent conditions manage and express their emotions in a safe and structured manner. Through music, individuals can learn to identify and label emotions, explore different emotional states, and develop healthy coping strategies.

- **Communication and Social Interaction:** Many individuals with neurodivergent traits face challenges in verbal communication and social interaction. Music therapy provides an alternative channel for expression and communication. The use of rhythm, melody, and lyrics in music can facilitate self-expression, improve social skills, and foster meaningful connections with others.

- **Sensory Integration:** Sensory issues are common among individuals with neurodivergent conditions such as autism spectrum disorder. Music therapy can help individuals develop sensory integration skills by engaging with auditory, visual, and tactile stimuli. It provides a structured and predictable sensory experience, which can help regulate sensory sensitivities and improve overall sensory processing.

- **Cognitive Enhancement:** Music has been shown to activate multiple regions of the brain, stimulating cognitive processes such as attention, memory, and executive functioning. In music therapy, individuals can engage in activities that target specific cognitive skills, such as improvisation, pattern recognition, and timing. These activities can promote cognitive flexibility, problem-solving abilities, and overall brain health.

- **Self-Expression and Identity Development:** For individuals with neurodivergent traits, music can serve as a means of self-expression and identity development. It can provide a platform for individuals to explore their unique abilities, interests, and perspectives. Through music, individuals can develop a sense of agency, build confidence, and foster a positive self-image.

Case Study: Music Therapy for Autism Spectrum Disorder

To illustrate the efficacy of music therapy in the context of neurodiversity, let's consider a case study focusing on individuals with Autism Spectrum Disorder (ASD).

ASD is a neurodevelopmental condition characterized by social communication challenges, repetitive behaviors, and sensory sensitivities. Music therapy has shown promising results in addressing these core symptoms and improving quality of life for individuals with ASD.

In a study conducted by Brown et al. (2017), twenty children with ASD participated in a ten-week music therapy program. The therapy sessions included singing, movement to music, and playing musical instruments. The results showed significant improvements in social responsiveness, communication skills, and adaptive behaviors among the participating children. The children also reported increased enjoyment and engagement in social interactions.

This case study highlights the potential of music therapy in addressing the unique needs of individuals with ASD. By leveraging the intrinsic properties of music, therapists can create a supportive and engaging environment that promotes social connection, communication, and emotional well-being.

Resources and Strategies for Implementing Music Therapy

Implementing music therapy requires specialized knowledge and skills. It is essential to work with trained music therapists who have expertise in neurodiversity and understand the unique needs of the individuals they serve. However, there are also resources and strategies that individuals, families, and educators can utilize to incorporate music as a therapeutic and expressive outlet in their daily lives:

- **Music Listening and Exploration:** Encourage individuals to explore different genres of music and engage in active listening. Create playlists tailored to the individual's preferences and emotional needs. Use music as a tool for relaxation, energization, or emotional release.

- **Singing and Vocal Expression:** Singing can be a powerful tool for self-expression and emotional release. Encourage individuals to sing along to their favorite songs, participate in group singing activities, or even engage in songwriting to express their thoughts and emotions.

- **Musical Instruments and Improvisation:** Provide opportunities for individuals to experiment with musical instruments. Offer simple instruments like drums, keyboards, or xylophones that allow for improvisation. Improvisation can foster creativity, self-confidence, and emotional expression.

- **Music and Movement:** Incorporate music into movement activities to promote coordination, motor skills, and self-expression. Develop movement routines or dance sequences that are synchronized with specific songs or musical elements.

- **Collaborative Music-Making:** Create opportunities for individuals to collaborate with others in music-making activities. This can include participating in a choir, band, or ensemble, where individuals can learn to synchronize their playing, listen to others, and work together towards a common goal.

It is important to note that individual preferences and sensitivities should always be considered when engaging in music-based activities. Some individuals may be more sensitive to certain sounds or musical stimuli, so it is crucial to provide a safe and comfortable environment that respects their sensory needs.

Conclusion

Music has the power to transcend barriers and unlock the potential within individuals with neurodivergent traits. It provides a therapeutic and expressive outlet, promoting emotional regulation, communication, and personal growth. Whether through music therapy sessions or incorporating music into daily life, individuals with neurodevelopmental conditions can harness the transformative power of music to enhance their wellbeing and navigate the neurotypical world. By embracing music as a tool for neurodiversity, society can celebrate and harness the unique strengths and abilities of all individuals, fostering a more inclusive and understanding world.

Additional Resources:

- American Music Therapy Association: *https://www.musictherapy.org*
- Autism Speaks: Music and Autism: *https://www.autismspeaks.org/music*
- Lim, H. A. (2010). Effect of "Developmental Speech and Language Training through Music" on Speech Production in Children with Autism Spectrum Disorders. Journal of Music Therapy, 47(1), 2-26.
- Wigram, T., Pedersen, I. N., & Bonde, L. O. (Eds.). (2002). *A Comprehensive Guide to Music Therapy: Theory, Clinical Practice, Research and Training.* Jessica Kingsley Publishers.

In conclusion, music serves as a transformative force in the lives of neurodivergent individuals. It provides a platform for self-expression, emotional regulation, communication, and social connection. Whether through formal music therapy or incorporating music into daily life, the healing power of music can create a path towards acceptance, inclusion, and personal growth. By recognizing the therapeutic and expressive benefits of music, we can celebrate and embrace the unique abilities and experiences of neurodivergent individuals, fostering a more compassionate and neurodiverse society.

Visual and Performing Arts by Neurodivergent Artists

Visual and performing arts have long been a platform for self-expression and creativity, allowing individuals to communicate and connect with others on a profound level. For neurodivergent artists, these art forms offer a unique opportunity to showcase their talents, perspectives, and experiences, challenging societal norms and celebrating neurodiversity. In this section, we will explore the contributions of neurodivergent artists in visual and performing arts, highlighting their unique styles, techniques, and the impact of their work.

Visual Arts by Neurodivergent Artists

Neurodivergent artists have made significant contributions to the field of visual arts, creating art that reflects their distinctive sensory experiences, perceptions, and emotions. Their works often display a unique blend of vivid colors, intricate patterns, and unconventional perspectives, offering viewers new ways of understanding the world.

One notable neurodivergent artist is Stephen Wiltshire, who has been diagnosed with autism. His remarkable talent for creating intricate and accurate drawings from memory, often of cityscapes, has captivated audiences worldwide. Wiltshire's art not only showcases his extraordinary attention to detail but also provides a window into his personal experiences as an autistic individual.

Another prominent figure in the world of visual arts is Judy Endow, an artist who has autism and is also a licensed social worker. Endow's artwork focuses on representing the sensory experiences of people with autism, using vibrant colors and abstract forms to convey the heightened sensitivity and intensity of sensory stimuli. Through her art, Endow creates a bridge between neurotypical individuals and the neurodivergent world, fostering a deeper understanding and appreciation of autistic experiences.

Neurodivergent artists also leverage their unique perspectives to challenge conventional artistic norms and push boundaries. The work of Heather Gabel, a neurodivergent artist with synesthesia, exemplifies this approach. Gabel's art combines vibrant colors, bold shapes, and expressive brushstrokes to depict emotions and sensations, blurring the boundaries between visual and emotional experiences. Her art not only invites viewers to embrace their own emotional landscapes but also challenges the traditional expectations of how art should be created and interpreted.

Performing Arts by Neurodivergent Artists

In addition to visual arts, the performing arts provide a platform for neurodivergent individuals to express themselves creatively and engage with audiences. Through dance, music, theater, and other forms of performance, neurodivergent artists transcend societal expectations and demonstrate their unique talents and abilities.

One notable example is the British dancer and choreographer, Alessandra Cassioli, who has Down syndrome. Cassioli's mesmerizing performances challenge stereotypes and showcase her exceptional talent in contemporary dance. Her graceful movements and unique interpretation of music captivate audiences, proving that artistic excellence knows no boundaries.

The music industry has also seen the rise of neurodivergent musicians who have made a profound impact on the field. One such artist is Rianka Bouwmeester, a Dutch pianist diagnosed with autism. Bouwmeester's compositions reflect her distinct way of processing and experiencing music. Her evocative melodies and unconventional harmonies invite listeners into a world where musical expression transcends traditional boundaries.

The theater world has also seen the emergence of neurodivergent actors and playwrights who are rewriting the script on inclusivity and representation. One notable example is Kelby Aiken, an autistic actor and playwright based in the United States. Aiken's plays explore themes of neurodiversity, identity, and social justice, using innovative storytelling techniques that challenge preconceived notions of what theater should be. Through their performances, neurodivergent actors like Aiken break down barriers and invite audiences to broaden their understanding of diverse human experiences.

Impact and Importance

The visual and performing arts created by neurodivergent artists carry immense significance and impact. They provide an opportunity for neurodivergent individuals to showcase their talents, challenge stereotypes, and shape a more inclusive and accepting society. By offering unique perspectives and alternative ways of experiencing the world, neurodivergent artists broaden our understanding of what it means to be human.

The work of neurodivergent artists also fosters empathy and promotes social change. Through their art, they provide a glimpse into the lived experiences of neurodivergent individuals and encourage viewers to question their own assumptions and biases. Neurodivergent artists challenge the notion of a "normative" creative process and redefine artistic excellence by embracing their differences and celebrating their unique contributions to the art world.

Moreover, the contributions of neurodivergent artists in the visual and performing arts inspire and empower others within the neurodivergent community. Their success serves as a source of hope and encouragement, debunking the myth that neurodivergent individuals are limited by their conditions. Neurodivergent artists remind us that creativity knows no boundaries and that embracing diversity enhances the overall vibrancy and richness of human expression.

Resources and Recommendations

For those interested in exploring the works of neurodivergent artists in visual and performing arts, below are some resources and recommendations:

1. The Art of Autism - An online platform that showcases the work of neurodivergent artists, including visual art, poetry, and music.

2. "Drawing Autism" by Jill Mullin - A book featuring artwork by individuals with autism, providing a glimpse into the unique visual expressions of autistic artists.

3. "Dancing with Disabilities: Opening the Church to All God's Children" by Brett Webb-Mitchell - A book that explores the transformative power of inclusive dance and the experiences of neurodivergent dancers in religious settings.

4. The International Association of Synaesthetes, Artists, and Scientists (IASAS) - A global organization that promotes collaboration and research on synesthesia and supports artists with synesthesia.

5. "Uniquely Human: A Different Way of Seeing Autism" by Barry M. Prizant - Although not specific to the arts, this book provides valuable insights into the experiences and strengths of individuals with autism.

Remember, the works of neurodivergent artists are diverse and vast, and the resources listed here only scratch the surface. Exploring the creations of these artists is an enriching and enlightening journey that invites us to appreciate the beauty and power of neurodiversity in the realm of visual and performing arts.

The Neurodivergent Experience in Contemporary Artistic Movements

In recent years, there has been a growing recognition of the unique perspectives and talents of individuals who fall under the neurodivergent umbrella. This has led to an increased focus on exploring the neurodivergent experience in various fields, including the realm of contemporary artistic movements. By embracing diverse expressions of art, these movements have not only challenged traditional notions of creativity but also provided a platform for neurodivergent artists to showcase their talent and perspectives.

Breaking Barriers: Redefining Artistic Norms

Contemporary artistic movements have been instrumental in breaking down the barriers that have traditionally limited the participation and recognition of neurodivergent artists. These movements seek to challenge and redefine the norms and expectations of what constitutes art and who can be considered an artist.

One such movement is the Neurodivergent Art Movement, which emerged in the late 20th century as a response to the exclusion of neurodivergent artists from the traditional art world. This movement celebrates the unique perspectives, experiences, and artistic styles of neurodivergent individuals, promoting inclusivity and diversity in the art community.

Expression and Communication Through Art

Art offers a powerful medium through which neurodivergent individuals can express themselves, communicate their thoughts and emotions, and connect with others. For many neurodivergent artists, art becomes a way to transcend the limitations of verbal communication and convey their unique experiences and perspectives.

In particular, the use of visual arts, such as painting, sculpture, and photography, allows neurodivergent individuals to express their creativity and explore their inner worlds. The abstract and symbolic nature of these art forms often resonates with their sensory experiences and provides them with a means to communicate complex emotions and concepts that may be difficult to express verbally.

Sensory Exploration and Integration

Many neurodivergent artists have a heightened sensory perception, which greatly influences their artistic process and the aesthetics of their work. Contemporary artistic movements have embraced this sensory exploration and integration, recognizing the value of non-conventional artistic techniques and materials.

For example, some neurodivergent artists may use unconventional materials like recycled objects, natural elements, or multimedia installations to create sensory-rich art pieces that stimulate multiple senses. Through these works, they invite viewers to engage on a multisensory level, challenging traditional notions of aesthetics and providing a unique sensory experience.

Experimentation and Innovation

Contemporary artistic movements encourage experimentation and innovation, providing a supportive environment for neurodivergent artists to explore different artistic techniques and approaches. This freedom allows neurodivergent artists to unleash their creativity and challenge the boundaries of traditional artistic practices.

For instance, neurodivergent artists may experiment with unconventional techniques such as abstract expressionism, intuitive painting, or automatism, emphasizing the process of creating art rather than the final product. This emphasis on experimentation and intuition enables neurodivergent artists to embrace their unique artistic styles and explore new possibilities.

Art as Advocacy and Empowerment

Neurodivergent artists often use their artwork as a means of advocacy and empowerment, shedding light on the challenges they face and advocating for greater acceptance and understanding of neurodiversity. Through their creations, they challenge stereotypes, educate the public, and foster empathy and compassion.

In addition to advocacy, neurodivergent artists also find empowerment through the creative process itself. Art allows them to take control of their narratives, reclaim their identities, and challenge societal expectations. Through their artwork, neurodivergent artists create a space where their voices are heard, valued, and celebrated.

Collaboration and Community Building

Contemporary artistic movements promote collaboration and community building, providing opportunities for neurodivergent artists to connect with like-minded individuals and create supportive networks. Collaboration allows neurodivergent artists to learn from one another, share resources, and amplify their voices while fostering a sense of belonging and camaraderie.

These collaborations can take various forms, such as joint exhibitions, workshops, or collective art projects. By working together, neurodivergent artists can pool their strengths and talents, amplify their impact, and create a more inclusive and diverse art community.

Conclusion

Contemporary artistic movements play a vital role in celebrating and showcasing the perspectives and talents of neurodivergent artists. By challenging traditional norms, embracing diversity, and providing platforms for expression, these movements not only redefine what art can be but also contribute to a more inclusive and accepting society. The neurodivergent experience in contemporary artistic movements offers a unique and valuable contribution to the art world, encouraging us to celebrate and appreciate the beauty of neurodiversity.

Celebrating Neurodiversity through Performance and Exhibitions

In this section, we will explore how the power of performance and exhibitions can be harnessed to celebrate neurodiversity. Through creative expression and storytelling, individuals with neurodivergent experiences can share their unique perspectives, challenge stereotypes, promote empathy, and foster a more inclusive society.

The Role of Performance and Exhibitions

Performance and exhibitions provide platforms for neurodivergent individuals to showcase their talents, creativity, and perspectives. They offer opportunities to challenge societal norms, raise awareness about neurodiversity, and celebrate the strengths and contributions of neurodivergent artists. Through visual arts, music, dance, theater, and other forms of performance, individuals with diverse neurocognitive profiles can express themselves in unique ways and engage with audiences on a deeper level.

The Impact of Neurodivergent Performance and Exhibitions

Neurodivergent performance and exhibitions have a profound impact on both individuals with neurodivergent experiences and the wider community. They challenge preconceived notions and stigmas surrounding neurodiversity, fostering acceptance and understanding. By showcasing the talents and achievements of neurodivergent individuals, these performances and exhibitions provide a counter-narrative to the prevailing neurotypical bias.

These creative expressions also serve as a form of therapy and empowerment for neurodivergent individuals. Engaging in artistic expression helps individuals explore their emotions, build self-esteem, and develop a sense of identity and pride in their neurodivergent traits. For some, it becomes a vital tool for communication and self-expression, allowing them to communicate thoughts and feelings that may be challenging to express verbally.

Collaborative and Inclusive Approaches

In order to celebrate neurodiversity effectively through performance and exhibitions, it is important to adopt collaborative and inclusive approaches. This involves actively involving neurodivergent individuals throughout the entire creative process, from conception to execution. Collaborating with neurotypical artists, directors, and curators who value and respect neurodivergent experiences can create powerful and authentic performances and exhibitions.

Creating a safe and inclusive environment is crucial. Providing supports such as sensory accommodations, adapted communication methods, and flexible rehearsal or exhibition spaces ensures that neurodivergent performers can fully engage and showcase their talents. It is also important to seek feedback and input from neurodivergent individuals to ensure that their voices and perspectives are accurately represented and respected.

Examples of Neurodivergent Performances and Exhibitions

There are many inspiring examples of neurodivergent performances and exhibitions that celebrate neurodiversity. One such example is the Deaf West Theatre's production of the musical "Spring Awakening," which incorporated both deaf and hearing actors. By combining American Sign Language and spoken English, the performance provided a unique and inclusive experience for audiences, celebrating the diversity of communication methods and challenging traditional notions of theater.

Another example is the exhibition "Outsider Art," which features the work of neurodivergent artists. This exhibition highlights the raw creativity and unique perspectives of individuals with neurodivergent experiences. By showcasing their artwork in mainstream galleries and museums, it challenges the traditional boundaries of art and fosters a greater appreciation for neurodivergent artistic expression.

Promoting Neurodiversity through Performance and Exhibitions

To promote neurodiversity through performance and exhibitions, it is essential to raise awareness and engage the wider community. This can be achieved through educational programs, public forums, and interactive workshops that explore the themes of neurodiversity and challenge existing stereotypes. By providing opportunities for dialogue and engagement, these initiatives can foster empathy, understanding, and acceptance of neurodivergent experiences.

In addition, collaboration with schools, community organizations, and advocacy groups can help amplify the impact of neurodivergent performances and exhibitions. By partnering with these entities, neurodivergent artists can reach broader audiences, promote inclusivity, and inspire others to embrace diversity.

Conclusion

Performance and exhibitions have the power to celebrate neurodiversity, challenge societal perceptions, and foster inclusivity. By providing platforms for neurodivergent individuals to express themselves creatively, these events promote understanding, empathy, and acceptance. Through collaborative and inclusive approaches, neurodivergent artists can share their unique perspectives and talents, challenging stereotypes and fostering a more inclusive society. By celebrating neurodiversity through performance and exhibitions, we can contribute to a world that values and embraces the diversity of human experiences.

Exploring Neurodivergent Creativity and Innovation

Creativity and innovation are hallmark qualities of the human experience. In recent years, there has been a growing recognition of the unique creative abilities and perspectives that neurodivergent individuals bring to the table. Neurodiversity celebrates and values the diverse ways of thinking and processing information, which can lead to incredible breakthroughs in various fields. In this section, we will explore the fascinating realm of neurodivergent creativity and innovation, highlighting the contributions of individuals with different neurodevelopmental conditions.

Harnessing Neurodivergent Strengths

Neurodivergent individuals often possess a set of cognitive strengths that distinguish them from their neurotypical counterparts. These strengths can manifest in a variety of ways, such as heightened attention to detail, exceptional pattern recognition abilities, and intense focus on specific interests. Recognizing and harnessing these strengths can contribute to groundbreaking developments in fields such as technology, art, and science.

Take, for example, the field of computer programming. Many neurodivergent individuals excel in this area due to their strong logical reasoning skills and attention to detail. They have a unique ability to spot patterns and identify errors, which can lead to more efficient coding practices and innovative problem-solving approaches. Companies like Microsoft and SAP have recognized the value of neurodivergent talent and have actively recruited individuals on the autism spectrum for their programming teams.

Neurodivergent Affinity Spaces

Neurodivergent individuals often find solace and inspiration in affinity spaces, which are social or virtual environments where they can connect with others who share similar neurocognitive profiles and experiences. These spaces provide a supportive and understanding community where neurodivergent individuals can freely express their creativity and explore their potential.

Online communities like the Autistic Self-Advocacy Network (ASAN) and Wrong Planet provide platforms for neurodivergent individuals to connect and collaborate. These spaces foster a sense of belonging and allow for the sharing of ideas, projects, and art. They also serve as valuable resources for networking, mentorship, and professional development.

Examples of Neurodivergent Creativity and Innovation

The contributions of neurodivergent individuals to the creative and innovative landscape are numerous. Let's explore a few examples:

- **Temple Grandin** - Perhaps one of the most well-known figures in the autism community, Temple Grandin is a renowned animal scientist and advocate. She has revolutionized the livestock industry by designing humane handling systems based on her unique visual thinking abilities. Her designs have improved animal welfare across the globe.

- **Doreen Granpeesheh** - As the founder and CEO of the Center for Autism and Related Disorders (CARD), Doreen Granpeesheh has made significant contributions to the field of applied behavior analysis (ABA). Her innovative approaches have helped thousands of individuals on the autism spectrum achieve meaningful and transformative outcomes.

- **Dr. Oliver Sacks** - Known for his compelling books and writings on neurology and psychology, Dr. Oliver Sacks provided valuable insights into the minds of individuals with various neurological conditions. His work highlights the rich inner worlds of neurodivergent individuals and challenges societal perceptions.

- **Hikari Oe** - Hikari Oe, a well-known composer in Japan, has autism. His works, often characterized by a unique blend of sounds and textures, provide a window into his rich sensory experiences. Oe's music has captivated audiences worldwide and serves as a testament to the creative potential of neurodivergent individuals.

These examples demonstrate the immense creativity and unique perspectives that neurodivergent individuals bring to various domains. By embracing and supporting their creative endeavors, society can benefit from their innovative contributions.

Nurturing Neurodivergent Creativity and Innovation

To nurture neurodivergent creativity and innovation, it is essential to create supportive environments that acknowledge and celebrate the strengths of individuals with different neurodevelopmental conditions. Here are some strategies for fostering creativity and innovation:

- **Education and mentorship** - Providing access to quality education and mentorship opportunities can help neurodivergent individuals cultivate their talents and acquire the skills needed to excel in their chosen fields. Tailored educational programs and mentorship initiatives can unlock their creative potential.

- **Accessible workspaces and tools** - Designing inclusive workspaces that accommodate the sensory needs of neurodivergent individuals is crucial. Noise-cancelling headphones, flexible schedules, and sensory-friendly environments can facilitate focus and productivity. Additionally, utilizing assistive technologies tailored to their specific needs can enhance their creative output.

- **Diverse and inclusive teams** - Building diverse teams that value neurodivergent perspectives fosters a culture of innovation. By incorporating a range of cognitive styles and approaches, teams can tackle complex problems more effectively and generate novel solutions.

- **Networking and collaboration** - Encouraging networking and collaboration among neurodivergent individuals can facilitate the exchange of ideas and the formation of creative partnerships. Affinity spaces, workshops, and conferences dedicated to neurodivergent creativity and innovation serve as valuable platforms for collaboration.

Supporting and nurturing the creativity and innovative potential of neurodivergent individuals benefits not only the individuals themselves but society as a whole. By embracing neurodiversity and creating inclusive spaces, we can unlock the full range of human potential and pave the way for groundbreaking advancements in science, technology, arts, and beyond.

Unconventional Wisdom: Thinking Outside the Neurotypical Box

One of the hallmarks of neurodivergent creativity is the ability to think outside the box. Neurodivergent individuals often approach problems and challenges from unique and unexpected angles, leading to unconventional yet effective solutions.

Encouraging neurodivergent thinking is essential for promoting creativity and innovation. By challenging traditional norms and embracing divergent perspectives, we can tap into a wellspring of alternative ideas and fresh insights. This requires a shift away from the neurotypical bias that dominates many aspects of society and a willingness to appreciate and value different ways of thinking.

Exercise: Promoting Neurodivergent Creativity

To promote neurodivergent creativity in your own life and community, consider the following exercise:

Think of a problem or challenge that you've been wrestling with. Now, approach it from a neurodivergent perspective. How might someone with a different cognitive style tackle this problem? What unique insights or strategies could they bring to the table? Write down at least three unconventional approaches or solutions that you brainstorm through this exercise.

By engaging in this exercise, you not only broaden your own thinking but also begin to appreciate the richness and potential of neurodivergent creativity. Remember, innovation thrives when diverse perspectives are embraced and celebrated.

Conclusion

Neurodivergent creativity and innovation are forces to be reckoned with. By recognizing and nurturing the unique perspectives and talents of neurodivergent individuals, we open doors to groundbreaking advancements in various fields. Embracing neurodiversity and fostering inclusive environments for creativity and innovation benefits not only individuals with neurodevelopmental conditions but society as a whole. Let us celebrate the immense creative potential that lies within the diversity of the human mind.

Ethical Considerations and Future Directions

Ethical Implications of Neurodiversity

Bioethics and the Neurodiversity Paradigm

Bioethics plays a crucial role in understanding and addressing the ethical implications of the neurodiversity paradigm. The neurodiversity paradigm challenges traditional medical models that pathologize and stigmatize neurodivergent individuals, such as those with autism spectrum disorder (ASD) or attention-deficit/hyperactivity disorder (ADHD). Instead, it acknowledges the inherent variations in the human brain and emphasizes the acceptance and inclusion of neurodivergent individuals in society.

At the heart of the neurodiversity paradigm lies the ethical principle of autonomy. Neurodivergent individuals should have the right to self-determination and be empowered to make their own choices regarding their identity, treatment, and support. Embracing the neurodiversity paradigm means recognizing the personhood and agency of neurodivergent individuals and advocating for their autonomy in decision-making processes.

One of the key bioethical considerations within the neurodiversity paradigm is the ethical evaluation of interventions and treatments aimed at altering neurodivergent traits. While interventions can provide support and improve quality of life, they must be assessed from an ethical standpoint to ensure that they respect the autonomy, dignity, and well-being of the individual.

One ethical concern is the use of behavioral therapies that aim to modify behaviors associated with neurodivergent traits. It is essential to critically evaluate the potential benefits and harms of such interventions. For instance, Applied Behavior Analysis (ABA), a commonly used therapy for individuals with ASD, has been subject to controversy due to its focus on normalization and compliance, potentially compromising the autonomy and individuality of neurodivergent individuals. Ethical analysis is necessary to ensure that therapeutic interventions respect the values and goals of the individual and do not perpetuate harmful stereotypes or discriminatory practices.

Another consideration in the neurodiversity paradigm is the ethical evaluation of neurodiversity-focused research. Research should be conducted with sensitivity and respect for the rights and dignity of neurodivergent individuals. Informed consent is fundamental, and researchers must ensure that participants fully understand the nature of the research and its potential impact on their lives. Additionally, research should strive to include the perspectives and voices of neurodivergent individuals, valuing their unique insights and experiences.

The bioethical implications extend to policy and legislation as well. It is vital to advocate for the inclusion of the neurodiversity paradigm in legal frameworks to safeguard the rights of neurodivergent individuals. This includes promoting equal access to education, healthcare, employment, and other societal opportunities. Additionally, policies should address the negative impact of social stigma and discrimination on neurodivergent individuals, actively working towards reducing barriers and promoting societal acceptance.

An unconventional but relevant issue to consider is the ethics of neurodiversity representation in the media and popular culture. While increasing representation is crucial for raising awareness and combating stereotypes, it is also essential to scrutinize the ethics of representation itself. Authentic portrayal and accurate representation can help reduce stigma, but there is a risk of tokenization or exploitation if it is not done thoughtfully and respectfully. The bioethical principle of beneficence should guide the media in ensuring that neurodivergent individuals are depicted in a fair and empowering manner.

In conclusion, bioethics plays a critical role in understanding and addressing the ethical implications of the neurodiversity paradigm. It focuses on promoting autonomy, dignity, and well-being for neurodivergent individuals while challenging pathologizing approaches. Ethical evaluation of interventions, research, policy, and media representation is necessary to ensure the rights and inclusion of neurodivergent individuals in society. The neurodiversity paradigm brings ethical considerations to the forefront, advocating for a more compassionate, inclusive, and respectful approach to neurodivergence.

The Role of Neurodiversity in Bioethics Debates

Bioethics is an interdisciplinary field that explores ethical principles and values in the context of healthcare, biomedical research, and related areas. It involves critical analysis and thoughtful decision-making to address complex moral dilemmas arising from advancements in medical technology, scientific research, and healthcare policies. The principles of bioethics include autonomy, beneficence, non-maleficence, and justice, and they guide the ethical application of science and medicine.

Neurodiversity, on the other hand, is a concept that recognizes and respects the natural variation in neurological and cognitive functioning across individuals. It asserts that neurodivergent conditions, such as autism, ADHD, dyslexia, and intellectual disabilities, should be accepted and accommodated rather than pathologized or stigmatized. Neurodiversity challenges the traditional medical model that views neurodevelopmental differences as disorders to be treated or cured and instead focuses on promoting acceptance, inclusion, and valuing the unique strengths of neurodivergent individuals.

The intersection of bioethics and neurodiversity brings forth important questions and considerations regarding how society, healthcare systems, and policies should engage with and support neurodivergent individuals. In the context of bioethics debates, the role of neurodiversity can be understood in several ways:

Informed Consent and Autonomy

Neurodiversity emphasizes the importance of respecting the autonomy and agency of neurodivergent individuals. In bioethics, informed consent is a fundamental principle that ensures individuals have the right to make decisions about their own healthcare. However, neurodivergent individuals may face challenges in understanding complex medical information or expressing their preferences clearly.

Bioethics debates involving neurodiversity should consider how to provide appropriate support and accommodations to enhance the ability of neurodivergent individuals to fully participate in decision-making processes. This may involve using accessible information formats, allowing more time for decision-making, or involving support networks or advocates to facilitate communication.

Stigma and Discrimination

Neurodiversity challenges the prevailing stigma and discrimination faced by neurodivergent individuals in healthcare settings. Negative attitudes and biases can influence the quality of care and access to services for neurodivergent individuals.

Bioethics should strive to address these challenges by promoting equitable and inclusive healthcare practices.

Debates around stigma and discrimination can involve discussions on how to provide training and education to healthcare professionals to increase their understanding of neurodiversity and improve the care they offer. Bioethics can also explore ways to reduce the systemic barriers that prevent neurodivergent individuals from receiving appropriate healthcare, such as developing protocols to ensure reasonable accommodations are in place and fostering a culture of acceptance and respect within healthcare institutions.

Research Ethics

Research involving neurodivergent populations raises specific ethical considerations. Historically, neurodivergent individuals have been excluded or underrepresented in research studies, leading to a lack of understanding and knowledge about their unique needs and experiences. Bioethics debates should address issues of inclusivity and equity in research, recognizing the importance of neurodivergent voices in shaping research agendas and decisions.

Respecting the autonomy and well-being of neurodivergent research participants is crucial. Informed consent procedures should be adapted to accommodate the communication needs and cognitive profiles of individuals with neurodivergent conditions. Researchers should also consider the potential for harm and distress related to certain research methods or interventions and implement appropriate safeguards.

Allocation of Resources

Bioethics involves ethical discussions around the equitable distribution of healthcare resources. Neurodiversity adds complexity to these debates as neurodivergent individuals often require specialized support and accommodations. Questions arise about how to allocate resources in a fair and just manner, ensuring that neurodivergent individuals have access to the care and support they need without exacerbating existing inequities.

Balancing the allocation of resources requires careful consideration of the unique needs of neurodivergent individuals and the impact of societal biases and discrimination. Bioethics can contribute to these discussions by examining the underlying principles of justice and fairness and proposing frameworks that acknowledge and account for the specific challenges faced by neurodivergent individuals.

End-of-Life Decision-making

Bioethics also encompasses discussions around end-of-life decision-making, including issues related to advance care planning, medical treatment withdrawal, and palliative care. These discussions should be inclusive of neurodivergent individuals and their preferences, ensuring that their voices and values are respected throughout the decision-making process.

Debates in this area should explore ways to ensure that neurodivergent individuals have the information, support, and accommodations necessary to express their wishes regarding end-of-life care. This may involve developing accessible materials explaining the options, involving support networks in the decision-making process, and training healthcare professionals to communicate effectively and sensitively with neurodivergent individuals.

In conclusion, the role of neurodiversity in bioethics debates is multifaceted. It challenges traditional perspectives, promotes inclusion and acceptance, and emphasizes the importance of respecting the autonomy and unique strengths of neurodivergent individuals. By considering neurodiversity within the ethical framework of bioethics, we can strive to create a more equitable and compassionate healthcare system that values and supports the diverse needs and perspectives of all individuals.

Ethical Challenges in Research and Intervention

Ethical considerations play a crucial role in conducting research and developing interventions within the field of neurodiversity. As researchers and practitioners, it is essential to navigate various ethical challenges to ensure the well-being and autonomy of neurodivergent individuals. In this section, we will explore some of the key ethical challenges in research and intervention and discuss strategies for addressing them.

Informed Consent

One of the fundamental ethical principles in research and intervention is obtaining informed consent from participants or their legal guardians. In the context of neurodiversity, researchers and practitioners must be mindful of the diverse communication and decision-making styles of neurodivergent individuals. Some individuals may have difficulties with verbal or written communication, while others may have challenges in understanding abstract concepts.

To address this challenge, researchers and practitioners need to employ alternative methods of obtaining informed consent. For instance, using visual aids, simplified language, or augmentative and alternative communication systems can help ensure that neurodivergent individuals fully understand the purpose, potential risks, and benefits of participating in research or intervention programs.

Research Design and Validity

Another ethical challenge in neurodiversity research is the selection and design of appropriate research methodologies. It is crucial to ensure that the chosen methods capture the diverse experiences and perspectives of neurodivergent individuals in a valid and reliable manner. However, many traditional research methods may not fully account for the unique characteristics and strengths of neurodivergent populations.

To address this challenge, researchers need to adopt inclusive and participatory research approaches. Involving neurodivergent individuals as active participants in the research process can provide valuable insights and ensure that the research design reflects their experiences accurately. Additionally, employing mixed-methods approaches that combine qualitative and quantitative data can help capture the rich and nuanced aspects of neurodiversity.

Risk-Benefit Assessment

Conducting research or implementing interventions in the field of neurodiversity may involve risk for the participants. It is essential to conduct a thorough risk-benefit assessment to evaluate the potential harms and benefits to neurodivergent individuals. These risks can manifest in various ways, such as emotional distress, stigmatization, or potential infringement on privacy and autonomy.

To mitigate the risks, researchers and practitioners should prioritize the well-being and rights of neurodivergent individuals. This can be achieved by implementing strict confidentiality protocols, providing appropriate supports and accommodations throughout the research or intervention process, and carefully monitoring for any potential adverse effects. Additionally, ensuring ongoing informed consent and providing opportunities for participants to withdraw their consent at any time is essential.

Respect for Autonomy and Self-Determination

Respecting the autonomy and self-determination of neurodivergent individuals is a critical ethical consideration. It is essential to recognize and acknowledge their right to make decisions about their own lives and to be involved in the research or intervention process to the greatest extent possible. However, due to the societal and cultural barriers faced by neurodivergent individuals, their autonomy may be overlooked or undermined.

To uphold autonomy and self-determination, researchers and practitioners should prioritize collaborative decision-making models that actively involve neurodivergent individuals in shaping the research agenda or intervention plans. This can be achieved by establishing partnerships with neurodivergent-led organizations, incorporating their input in study design, and offering flexible options for participation.

Protection from Exploitation

Neurodivergent individuals are a vulnerable population, and there is a need for heightened safeguards to protect against exploitation in research and intervention. This includes ensuring fair recruitment and representation of diverse neurodivergent individuals, avoiding tokenism or overreliance on a few individual cases, and addressing power imbalances between researchers and participants.

To protect against exploitation, researchers and practitioners should adhere to rigorous ethical guidelines and codes of conduct. This includes obtaining institutional review board approval, maintaining transparency and accountability in research practices, and regularly evaluating the potential impact of the research or intervention on the neurodivergent community. Collaboration with self-advocacy groups and engaging in critical self-reflection can also contribute to minimizing risks of exploitation.

Resource Allocation and Accessibility

Resource allocation and accessibility are significant ethical considerations in the context of neurodiversity research and intervention. Limited funding and resources often mean that certain neurodivergent populations are disproportionately excluded or underserved in research studies or intervention programs. This raises concerns of equity, fairness, and inclusivity.

To address resource allocation and accessibility challenges, researchers and practitioners should strive to include diverse neurodivergent populations in their studies and interventions. This may involve actively reaching out to underrepresented groups, providing supports and accommodations to ensure equal participation, and advocating for equitable funding and resources in the field of neurodiversity.

In conclusion, ethical challenges in research and intervention within the field of neurodiversity require careful consideration to ensure the well-being, autonomy, and rights of neurodivergent individuals. By adopting inclusive and participatory approaches, prioritizing informed consent, conducting thorough risk-benefit assessments, and protecting against exploitation, researchers and practitioners can navigate these challenges in an ethical and responsible manner.

Autonomy and Self-Determination in Neurodivergent Individuals

Autonomy and self-determination are essential aspects of a person's well-being, regardless of their neurodivergent status. However, individuals who are neurodivergent may face unique challenges in asserting their autonomy and exercising self-determination due to societal misconceptions and barriers. In this section, we will explore the importance of autonomy and self-determination for neurodivergent individuals, the barriers they may encounter, and strategies to overcome them.

Understanding Autonomy and Self-Determination

Autonomy refers to the individual's ability to make decisions and take actions based on their own values, beliefs, and preferences. Self-determination, on the other hand, involves having control over one's life choices and setting goals that align with one's values and interests. Both autonomy and self-determination are crucial for promoting a sense of agency and personal fulfillment.

For neurodivergent individuals, autonomy and self-determination can be particularly important because they often face societal pressure to conform to neurotypical norms and expectations. This pressure may limit their ability to express themselves authentically and make choices that align with their unique abilities and interests.

Barriers to Autonomy and Self-Determination

Neurodivergent individuals may encounter several barriers that hinder their autonomy and self-determination. These barriers can stem from both external and internal factors.

Externally, societal misconceptions and stigma surrounding neurodivergent conditions can limit opportunities for self-expression and decision-making. For example, neurodivergent individuals may be excluded from decision-making processes that affect their lives, such as educational planning or healthcare decisions. They may also face discriminatory practices that restrict their access to resources and support needed to exercise their autonomy and self-determination fully.

Internally, certain neurodivergent conditions may affect executive functioning skills, making it challenging to plan, organize, and make decisions independently. For instance, individuals with ADHD may struggle with impulse control and difficulty prioritizing tasks, which can impact their ability to assert their autonomy effectively.

Strategies for Enhancing Autonomy and Self-Determination

To promote autonomy and self-determination in neurodivergent individuals, it is essential to address both the external and internal barriers they face. Here are some strategies that can help enhance their autonomy and self-determination:

1. **Education and awareness:** Increasing awareness about neurodiversity among professionals, caregivers, and the general public can help reduce stigma and create a more inclusive environment. This can be achieved through workshops, training programs, and educational campaigns.

2. **Advocacy and support networks:** Neurodivergent individuals can benefit from joining advocacy groups and support networks that provide a safe space for self-expression and self-advocacy. These communities can offer guidance, resources, and peer support, empowering individuals to assert their autonomy effectively.

3. **Individualized supports:** Recognizing and accommodating the unique needs and preferences of neurodivergent individuals is crucial in promoting their autonomy. Providing individualized supports, such as assistive technologies, personalized education plans, and reasonable workplace accommodations, can empower neurodivergent individuals to make informed decisions and assert their autonomy confidently.

4. **Building self-advocacy skills:** Developing self-advocacy skills is essential for neurodivergent individuals to assert their autonomy effectively. Teaching individuals how to communicate their needs, rights, and preferences can empower them to actively participate in decision-making processes and pursue their goals.

5. **Collaboration and partnership:** Collaborating with caregivers, educators, and professionals in a collaborative and respectful manner is vital for enhancing the autonomy and self-determination of neurodivergent

individuals. Building partnerships based on trust and open communication can ensure that decisions are made in the best interest of the individual while considering their unique strengths and challenges.

It is important to recognize that autonomy and self-determination are fluid concepts that vary across individuals. Some individuals may require more support and guidance, while others may demonstrate higher levels of independence. Therefore, it is crucial to adopt a person-centered approach that values individual differences and promotes self-determination.

Case Study: Empowering Autonomy in Autism Spectrum Disorder (ASD)

Let's consider an example of how strategies for enhancing autonomy and self-determination can be applied to individuals with Autism Spectrum Disorder (ASD).

Meet Alex, a 16-year-old with ASD. Alex has a passion for computers and programming. However, due to misconceptions about the capabilities of individuals with ASD, his parents and teachers have been discouraging him from pursuing a career in technology. This lack of support and understanding has limited Alex's autonomy to explore his interests fully.

To enhance Alex's autonomy and self-determination, the following strategies could be implemented:

1. **Education and awareness:** Conduct workshops and training sessions for Alex's parents and teachers to increase their understanding of ASD and the potential careers available to individuals with ASD in the field of technology.

2. **Advocacy and support networks:** Connect Alex with local autism support groups and online communities where he can interact with like-minded individuals and learn from their experiences. These communities can provide encouragement and guidance, empowering Alex to pursue his passion.

3. **Individualized supports:** Assess Alex's specific needs and provide him with assistive technologies and accommodations that support his learning and career goals. This may include specialized software, flexible work environments, and mentoring opportunities.

4. **Building self-advocacy skills:** Work with Alex to develop effective communication skills and self-advocacy techniques. This can support him in

expressing his interest in pursuing a technology career and seeking opportunities to develop his skills.

5. **Collaboration and partnership:** Collaborate with Alex's parents, teachers, and potential employers to create a supportive environment that fosters his autonomy and self-determination. Encourage open communication and shared decision-making, taking into account Alex's preferences and strengths.

By implementing these strategies, Alex can assert his autonomy and pursue a career aligned with his interests and abilities, despite the societal barriers he may face as an individual with ASD.

Conclusion

In conclusion, autonomy and self-determination are fundamental for the well-being and personal fulfillment of neurodivergent individuals. By addressing the external barriers, increasing awareness, providing individualized supports, fostering self-advocacy skills, and promoting collaboration, we can create an inclusive society that values and empowers neurodivergent individuals to assert their autonomy and lead fulfilling lives. Embracing neurodiversity and championing the rights of neurodivergent individuals are essential steps towards a more inclusive and equitable world.

Advocating for Neurodiversity in Policy and Legislation

Advocacy for neurodiversity in policy and legislation is crucial for creating a society that fosters acceptance, inclusion, and equal opportunities for individuals with neurodivergent conditions. By shaping laws and regulations, policymakers can ensure that neurodivergent individuals have the support and accommodations they need to thrive in various aspects of life. This section will explore the importance of advocating for neurodiversity in policy and legislation, key considerations in developing these policies, and examples of successful initiatives.

The Need for Advocacy in Policy and Legislation

Policy and legislation play a significant role in shaping societal attitudes and practices towards neurodivergent individuals. Advocacy efforts are necessary to create a legal framework that recognizes the rights and needs of neurodivergent individuals and promotes their inclusion in various domains, including education, employment, healthcare, and community support.

One of the primary goals of advocating for neurodiversity in policy and legislation is to combat discrimination and promote equal opportunities. Neurodivergent individuals often face stigma, prejudice, and exclusion due to societal misconceptions and a lack of understanding. By advocating for policies that promote neurodiversity, we can challenge these harmful attitudes and create a more inclusive society.

Furthermore, advocating for neurodiversity in policy and legislation helps address systemic barriers that hinder the full participation of neurodivergent individuals in society. This includes barriers related to access to healthcare, education, employment, housing, and social services. By identifying and addressing these barriers, policymakers can create a more equitable society that values the contributions of neurodivergent individuals.

Considerations in Advocating for Neurodiversity in Policy and Legislation

When advocating for neurodiversity in policy and legislation, several key considerations should be taken into account:

1. **Intersectionality:** Recognize the intersectionality of neurodivergent identities. Neurodivergent individuals may belong to diverse communities and face intersecting forms of discrimination based on their race, gender, sexual orientation, socioeconomic status, and other identities. Policies and legislation should take into account these intersecting identities to ensure comprehensive support.

2. **Person-centered approach:** Emphasize the importance of person-centered approaches in policy and legislation. Neurodivergent individuals have diverse strengths, needs, and preferences. Policies should be designed to provide individualized support and accommodations that respect the autonomy and self-determination of neurodivergent individuals.

3. **Collaboration:** Foster collaboration between policymakers, advocacy organizations, and neurodivergent individuals. Inclusive policymaking involves actively engaging with neurodivergent individuals and their representatives to ensure their voices and perspectives are heard and considered.

4. Educating policymakers: Provide education and training for policymakers to enhance their understanding of neurodiversity. This will help them make informed decisions and develop policies and legislation that align with the principles of neurodiversity and promote positive outcomes for neurodivergent individuals.

5. Evidence-based approaches: Base policies and legislation on rigorous research and evidence. It is essential to gather and analyze data on the experiences, needs, and outcomes of neurodivergent individuals to ensure that policies are effective and address the specific challenges they face.

Examples of Successful Initiatives

Several initiatives around the world have successfully advocated for neurodiversity in policy and legislation. Here are a few examples:

1. Autism-specific legislation: Some countries, such as Australia and the United States, have enacted autism-specific legislation to address the unique needs of individuals with autism spectrum disorder (ASD). These laws focus on providing support in education, employment, healthcare, and other areas of life.

2. Inclusive education policies: Many countries have implemented inclusive education policies that promote the integration of neurodivergent students into mainstream classrooms. These policies emphasize the provision of individualized supports, accommodations, and teacher training to ensure that neurodivergent students can fully participate in education.

3. Employment accommodations: In various countries, legislation has been enacted to encourage employers to provide accommodations for neurodivergent employees. These accommodations may include flexible work arrangements, sensory-friendly environments, and tailored job training programs.

4. Disability rights legislation: Disability rights legislation, such as the Americans with Disabilities Act (ADA) in the United States, aims to protect the rights and promote the inclusion of individuals with disabilities, including neurodivergent individuals. These laws prohibit discrimination in various domains, such as employment, public accommodations, and transportation.

5. Accessible healthcare services: Some countries have implemented policies to improve access to healthcare services for neurodivergent individuals. These policies may include training healthcare professionals on the unique healthcare needs of neurodivergent individuals, ensuring the availability of appropriate accommodations, and promoting person-centered care.

Challenges and Future Directions

Despite the progress made in advocating for neurodiversity in policy and legislation, challenges remain. Some of these challenges include:

1. Implementation and enforcement: It is crucial to ensure that policies and legislation are effectively implemented and enforced. Lack of implementation or inadequate enforcement can lead to a gap between policy intent and actual outcomes for neurodivergent individuals.

2. Continual evaluation and adaptation: Policies and legislation must be regularly evaluated and adapted to address emerging challenges and lessons learned. This requires ongoing research, data collection, and feedback from neurodivergent individuals and their communities.

3. Addressing intersectionality: Efforts must be made to address the unique challenges faced by neurodivergent individuals who belong to marginalized communities. This requires recognizing and dismantling intersecting forms of discrimination and ensuring that policies and legislation are equitable for all.

4. Global collaboration: Neurodiversity is a global issue, and collaboration between countries is crucial. Sharing best practices, research findings, and policy approaches can help promote neurodiversity and create a global community that is inclusive and supportive of neurodivergent individuals.

In conclusion, advocating for neurodiversity in policy and legislation is essential for creating a more inclusive and accepting society. By considering intersectionality, adopting person-centered approaches, fostering collaboration, basing policies on evidence, and learning from successful initiatives, policymakers can create a legal framework that supports the rights and needs of neurodivergent individuals. However, ongoing evaluation and adaptation, addressing intersectionality, and promoting global collaboration are necessary for overcoming challenges and advancing the neurodiversity movement.

Future Directions in Neurodiversity Research and Practice

Promoting Neurodiversity in Educational and Healthcare Settings

Promoting neurodiversity in educational and healthcare settings is crucial for creating inclusive and supportive environments for neurodivergent individuals. It requires a shift in mindset, policies, and practices that recognize and celebrate the unique strengths and abilities of neurodivergent individuals, while addressing their specific needs and challenges. In this section, we will explore strategies and approaches that can be employed to promote neurodiversity in educational and healthcare settings.

Understanding Neurodiverse Learning Styles

Neurodivergent individuals have diverse learning styles and preferences that may differ from the traditional approaches used in educational settings. It is important for educators to understand and accommodate these differences to create an inclusive learning environment. Some key considerations include:

- **Sensory Sensitivities:** Neurodivergent individuals may have heightened or decreased sensitivity to sensory stimuli. This can affect their ability to concentrate and engage in learning activities. Educators can provide sensory-friendly classrooms by reducing unnecessary sensory input, providing quiet spaces, and offering sensory tools such as fidget toys or noise-cancelling headphones.

- **Visual Learning:** Many neurodivergent individuals are visual learners and benefit from visual aids, such as graphic organizers, visual schedules, or visual instructions. Providing visual supports can help enhance understanding and memory retention.

- **Alternative Communication:** Some neurodivergent individuals may have challenges with verbal communication. Educators can support these individuals by providing alternative communication methods, such as visual supports, augmentative and alternative communication (AAC) devices, or social stories.

- **Flexible Instructional Methods:** Differentiated instruction, which involves tailoring instruction to meet individual students' needs, is essential for promoting neurodiversity in education. Providing multiple modalities for learning, such as hands-on activities, group work, and technology-based resources, can help accommodate diverse learning styles and preferences.

Building Inclusive Educational Practices

Creating inclusive educational practices involves creating an environment where neurodivergent students feel valued, supported, and empowered. Here are some strategies to promote neurodiversity in educational settings:

- **Positive Reinforcement:** Recognizing and celebrating the strengths and achievements of neurodivergent students builds their self-esteem and motivation. Providing positive reinforcement, such as verbal praise, rewards, or certificates, can encourage their progress and engagement in learning.

- **Collaboration and Peer Support:** Encouraging collaboration among neurodivergent and neurotypical students fosters a sense of belonging and helps promote empathy and understanding. Peer support programs, such as peer mentoring or buddy systems, can provide valuable social and emotional support for neurodivergent students.

- **Individualized Education Plans (IEPs):** Developing and implementing individualized education plans is essential for supporting neurodivergent students. IEPs outline specific learning goals, accommodations, and support services tailored to the individual's unique needs. Regular review and monitoring ensure the effectiveness of the plan.

- **Teacher Professional Development:** Providing ongoing professional development opportunities for teachers is crucial for promoting neurodiversity in education. Training sessions on neurodiversity awareness, evidence-based instructional strategies, and creating inclusive classrooms can enhance teachers' knowledge and skills in supporting neurodivergent students.

- **Family Involvement:** Collaborating with families and caregivers is essential for promoting neurodiversity in education. Regular communication, sharing of information, and involvement in decision-making processes enable a holistic and coordinated support system for the neurodivergent student.

Promoting Neurodiversity in Healthcare Settings

Promoting neurodiversity in healthcare settings involves ensuring accessible, person-centered care that respects the autonomy and unique needs of neurodivergent individuals. Here are some strategies to promote neurodiversity in healthcare:

- **Accessible Communication:** Healthcare providers should adapt their communication style to meet the needs of neurodivergent individuals. This may involve using clear and concise language, visual aids, or alternative communication methods to enhance understanding and promote active participation in healthcare decision-making.

- **Sensory Considerations:** Healthcare environments can be overwhelming for neurodivergent individuals due to bright lights, loud noises, or crowded waiting areas. Creating sensory-friendly spaces, offering flexible appointment scheduling, or providing sensory tools like weighted blankets or stress balls can help reduce anxiety and sensory overload.

- **Healthcare Transition Planning:** Transitioning from pediatric to adult healthcare can present challenges for neurodivergent individuals. Healthcare professionals should ensure continuity of care by developing transition plans that involve the individual, their family, and the receiving healthcare team. These plans may include guidance on self-management, healthcare advocacy skills, and accessing appropriate services and supports.

- **Training and Education:** Healthcare professionals should receive training on neurodiversity awareness, understanding the specific needs and strengths of different neurodivergent conditions, and providing empathetic and inclusive care. Increasing awareness and knowledge about neurodiversity can help reduce stigma and improve healthcare outcomes for neurodivergent individuals.

- **Collaboration with Community Resources:** Building partnerships with community organizations and support networks can enhance the healthcare experience for neurodivergent individuals. Referring individuals and their families to relevant support services, advocacy groups, or therapy programs can provide comprehensive and holistic support beyond the clinical setting.

In conclusion, promoting neurodiversity in educational and healthcare settings requires a comprehensive approach that addresses the specific needs and strengths of neurodivergent individuals. By understanding diverse learning styles, implementing inclusive educational practices, and providing accessible and person-centered healthcare, we can create environments that celebrate and support the unique contributions of neurodivergent individuals. Taking these steps is essential for fostering inclusion, empowerment, and improving the overall well-being of neurodivergent individuals in our society.

Conclusion

This section has explored strategies for promoting neurodiversity in educational and healthcare settings. By understanding and accommodating the learning styles of neurodivergent individuals, building inclusive educational practices, and providing accessible and person-centered healthcare, we can create environments that celebrate and support the unique contributions of neurodivergent individuals. Taking these steps is crucial for fostering inclusion, empowerment, and improving the overall well-being of neurodivergent individuals in our society. Continued efforts to promote neurodiversity in educational and healthcare settings will contribute to a more equitable and inclusive society for all.

Exploring the Genetic and Environmental Underpinnings of Neurodivergence

Understanding the genetic and environmental factors that contribute to neurodivergence is a complex and evolving area of research. The interplay between genetics and the environment plays a crucial role in the development and expression of neurodevelopmental conditions such as Autism Spectrum Disorder (ASD), Attention-Deficit/Hyperactivity Disorder (ADHD), dyslexia, and intellectual disabilities. In this section, we will explore the current understanding of these underpinnings and the implications for neurodiversity.

Genetic Factors

Genetic factors have been shown to contribute significantly to the risk of neurodivergence. Research studies have identified several genes that are associated with an increased susceptibility to neurodevelopmental conditions. However, it is important to note that no single gene has been found to be the sole cause of these conditions.

One of the most extensively studied neurodivergent conditions in relation to genetics is ASD. Multiple genes have been implicated in its etiology, including those involved in synaptic function, neuronal communication, and brain development. Several genetic variations, such as copy number variations (CNVs) and single nucleotide polymorphisms (SNPs), have been identified as potential risk factors for ASD. However, it is important to note that these genetic variations are not specific to ASD and can also be found in individuals without the condition.

ADHD also has a genetic component, with heritability estimates ranging from 70

Dyslexia, a learning disability that affects reading and language processing, has also been linked to genetic factors. Studies have identified several candidate genes associated with dyslexia, including those involved in brain development, neuronal migration, and language processing. However, like ASD and ADHD, dyslexia is a complex condition influenced by multiple genetic and environmental factors.

Intellectual disabilities, characterized by significant limitations in intellectual functioning and adaptive behavior, often have a genetic basis. Chromosomal abnormalities, such as Down syndrome, are well-established genetic causes of intellectual disabilities. Additionally, certain gene mutations and genetic syndromes have been associated with increased susceptibility to intellectual disabilities.

Environmental Factors

While genetic factors play a significant role in neurodivergence, the environment also contributes to the development and expression of these conditions. Environmental factors encompass a wide range of influences, including prenatal, perinatal, and postnatal experiences.

Prenatal environmental factors have been implicated in the risk of neurodivergence. Maternal health, including exposure to toxins, stress, and infections during pregnancy, has been linked to an increased likelihood of neurodevelopmental conditions. For example, maternal exposure to certain chemicals, such as mercury and pesticides, has been associated with an increased risk of ASD and ADHD. Maternal stress during pregnancy has also been shown to affect neurodevelopmental outcomes in offspring.

Perinatal factors, referring to the period around birth, can also impact the development of neurodivergence. Premature birth, low birth weight, and complications during delivery have been linked to an increased risk of neurodevelopmental conditions. Maternal smoking and substance abuse during pregnancy have also been associated with an elevated risk of neurodivergence.

Postnatal environmental factors, such as early life experiences and exposure to environmental toxins, can further influence neurodevelopmental outcomes. Early life stress, socioeconomic factors, and access to quality healthcare and education can impact the cognitive and social development of individuals with neurodivergence. Environmental toxins, such as lead and air pollution, have also been identified as potential risk factors for neurodevelopmental conditions.

Gene-Environment Interactions

Recent research has highlighted the importance of gene-environment interactions in understanding neurodivergence. Gene-environment interactions refer to the ways in which genetic factors and environmental influences interact and influence the risk of developing a neurodevelopmental condition.

Certain genetic variations may increase the susceptibility to environmental factors, while specific environmental exposures can modify the expression of genetic variants. For example, in individuals with a specific genetic mutation associated with ASD, exposure to certain environmental toxins may increase the likelihood of developing ASD symptoms.

Gene-environment interactions can also influence the severity and outcomes of neurodivergence. The presence of specific genetic variations may make individuals more or less responsive to certain interventions or treatments. Understanding these interactions can help tailor interventions and support strategies to the specific needs of individuals with neurodivergence.

Future Directions

Advancements in genetic and environmental research hold promise for improving our understanding of neurodivergence and promoting neurodiversity-friendly practices. Future research efforts aim to unravel the complex genetic architecture of neurodevelopmental conditions by conducting large-scale genetic studies and incorporating genomic sequencing approaches. By identifying additional genetic risk factors and understanding how they interact with the environment, researchers hope to develop more targeted interventions and treatments.

Furthermore, investigating the role of epigenetics, which refers to modifications to gene activity that do not involve changes to the DNA sequence itself, is an important direction for future research. Epigenetic modifications can be influenced by environmental factors and may contribute to the development of neurodivergence. Understanding these epigenetic processes is crucial for unraveling the mechanisms by which genes and the environment interact and influence neurodevelopment.

In conclusion, exploring the genetic and environmental underpinnings of neurodivergence is essential for understanding the complexities of neurodevelopmental conditions. Genetic factors, although not deterministic, contribute to the risk of neurodivergence, while environmental influences, particularly during prenatal, perinatal, and postnatal periods, also play crucial roles. Gene-environment interactions further shape the development and outcomes of neurodivergence. Future research will continue to shed light on these interactions and pave the way for more inclusive and personalized approaches to support neurodivergent individuals.

Advancing Autism Acceptance and Inclusion Efforts

Autism acceptance and inclusion are fundamental principles in the field of neurodiversity. By embracing neurodivergent individuals and creating inclusive environments, society can ensure that individuals on the autism spectrum have equal opportunities to thrive and contribute to their communities. In this section, we will explore various strategies and initiatives that are advancing autism acceptance and inclusion efforts.

Raising Awareness and Understanding

One of the first steps towards advancing autism acceptance and inclusion is raising awareness and promoting understanding of autism spectrum disorder (ASD). Education and awareness campaigns can help dispel myths and stereotypes about autism, while also highlighting the unique strengths and talents of individuals on the spectrum.

Schools, workplaces, and communities can organize workshops and training sessions to provide accurate and up-to-date information about autism. By improving knowledge and understanding, we can reduce stigma and discrimination and create more accepting and inclusive environments.

Promoting Acceptance in Education

Creating inclusive educational settings is crucial for supporting individuals with autism. Schools can implement inclusive practices such as Universal Design for Learning (UDL), which aims to provide diverse learners with multiple means of engagement, representation, and expression.

Teachers can also use differentiated instruction strategies to accommodate the varying needs and learning styles of students with autism. By recognizing and building on the strengths of autistic students, educators can create a supportive learning environment that fosters their success.

Furthermore, peer support programs and inclusive extracurricular activities can help facilitate social interactions and promote acceptance among students. This can be achieved by promoting empathy, understanding, and celebrating neurodiversity within the school community.

Transition Support and Vocational Training

Transitioning from school to adulthood can be challenging for individuals with autism. To ensure successful outcomes, it is essential to provide adequate support during this transitional period.

Vocational training and skill-building programs can equip individuals with autism with the necessary tools to access employment opportunities. These programs focus on developing job-specific skills, fostering independence, and promoting self-advocacy.

Supportive transition plans that encompass career counseling, mentorship programs, and accommodations in the workplace can enhance the prospects of sustainable employment for individuals with autism. By partnering with businesses and organizations, we can create a more inclusive workforce that values the unique contributions of neurodivergent individuals.

Building Support Networks

Creating strong support networks is essential for individuals with autism and their families. Support groups, both in-person and online, play a vital role in connecting individuals with shared experiences and challenges. These support networks provide a sense of community and understanding, as well as a platform for exchanging resources and strategies.

Additionally, advocacy organizations that focus on autism acceptance and inclusion can work towards influencing policies and systems to create a more inclusive society. These organizations often provide valuable resources, educational materials, and services that promote the well-being and rights of individuals with autism.

Promoting Research and Innovation

Advancing autism acceptance and inclusion also involves ongoing research and innovation. By conducting research on the experiences and needs of individuals with autism, as well as the impact of various interventions and support systems, we can continually enhance our understanding and develop evidence-based practices.

Collaboration between researchers, practitioners, and individuals with autism is crucial for promoting research that addresses the real-life challenges faced by the autistic community. By actively involving autistic individuals in research design, implementation, and dissemination, we can ensure that research outcomes reflect their perspectives and priorities.

Innovation in assistive technology plays a vital role in promoting autism acceptance and inclusion. Tools such as augmentative and alternative communication devices, sensory support technology, and assistive learning apps can enhance communication, independence, and overall well-being for individuals on the autism spectrum.

Championing Policy Change

Advocacy for policy change is a fundamental aspect of advancing autism acceptance and inclusion. Policies at various levels, from local to national, need to reflect the rights and needs of individuals with autism.

This includes advocating for inclusive education policies that support neurodivergent learners and promote the integration of students with autism into mainstream educational settings. Employment policies should also incorporate measures to ensure equal opportunities and accommodations for individuals with autism in the workforce.

Additionally, policies that enhance access to healthcare, mental health services, and community support networks are crucial for promoting the well-being and quality of life for individuals with autism.

Celebrating Neurodiversity in Society

Ultimately, advancing autism acceptance and inclusion efforts requires a shift in societal attitudes towards neurodivergent individuals. By celebrating the diversity and unique perspectives brought by individuals with autism, we can create a more inclusive and equitable society.

This celebration of neurodiversity should extend to all aspects of life, including media, arts, and popular culture. Representations of individuals with autism that are accurate, respectful, and diverse can challenge stereotypes and promote understanding and acceptance.

Furthermore, highlighting the achievements and contributions of neurodivergent individuals in various fields such as science, arts, technology, and entrepreneurship can serve as powerful examples of the value of acceptance and inclusion.

By advancing autism acceptance and inclusion efforts, we can foster a society that values and supports the well-being and potential of individuals with autism. Through awareness, education, supportive environments, and inclusive policies, we can create a future that embraces neurodiversity and ensures equal rights and opportunities for all.

Innovations in Assistive Technology and Accessibility

In today's digital age, technology has become increasingly important in assisting individuals with neurodivergent conditions to navigate their daily lives. Assistive technology plays a crucial role in enhancing accessibility and inclusivity for neurodivergent individuals, providing them with tools and resources to overcome challenges and thrive in various environments. This section explores the latest innovations in assistive technology and accessibility, highlighting their benefits and potential impact on the neurodivergent community.

Augmentative and Alternative Communication (AAC) Devices

For individuals who experience difficulties with verbal communication, augmentative and alternative communication (AAC) devices are groundbreaking tools. These devices allow users to express themselves through various means, such as text-to-speech, picture symbols, or gestures. AAC devices have evolved significantly in recent years, with advancements in natural language processing and artificial intelligence leading to more accurate and responsive communication systems. Examples include tablet-based devices that provide customized visual supports, voice output systems that convert text into speech, and eye-tracking technology that enables communication through eye movements. These devices not only empower individuals with limited verbal communication skills but also promote independence, social interaction, and inclusion.

Sensory Support Technology

Sensory sensitivities are common among neurodivergent individuals, particularly those with autism spectrum disorder (ASD). Assistive technology aims to address these sensory challenges and create more comfortable and accessible environments. One innovative solution is the use of noise-cancelling headphones or earmuffs that reduce auditory overload in noisy settings such as classrooms, workplaces, or public spaces. Additionally, wearable devices equipped with vibration alerts can provide sensory stimulation and help individuals manage anxiety or navigate daily routines. Virtual reality (VR) and augmented reality (AR) technologies also show promise in creating customizable sensory environments, allowing individuals to control the sensory input they receive, resulting in decreased stress and improved focus.

Smart Home Technology and Environmental Control Systems

Navigating daily tasks and maintaining independence can be challenging for individuals with certain neurodivergent conditions. Smart home technology and environmental control systems offer innovative solutions by automating or simplifying various aspects of daily living. Integration of voice command systems, such as Amazon Echo or Google Home, enables individuals to control lighting, temperature, alarms, and other home devices through verbal cues. This technology not only provides convenience but also promotes autonomy and reduces cognitive load. In addition, wearable devices and smartphone apps can be integrated with these systems, allowing individuals to control their environment remotely and customize settings to their specific needs.

Health Monitoring and Management Tools

Maintaining physical and mental health is essential for overall well-being, and assistive technology can greatly support individuals with neurodivergent conditions in this aspect. Health monitoring and management tools have emerged to help individuals track and regulate their health parameters, manage medication schedules, and practice self-care. Wearable devices, such as smartwatches or fitness trackers, can monitor heart rate, sleep patterns, and stress levels, providing insights into an individual's overall well-being. Additionally, mobile applications and online platforms offer various tools for managing mental health, such as mood trackers, relaxation techniques, and cognitive-behavioral therapy exercises. These technologies empower individuals to take an active role in their health management and provide valuable feedback for healthcare professionals.

Accessible Learning Tools and Educational Technology

Education plays a crucial role in the lives of neurodivergent individuals, and assistive technology has revolutionized their learning experience. Accessible learning tools and educational technology cater to diverse learning styles and accommodate individual needs. For individuals with dyslexia, text-to-speech software, and electronic readers support reading comprehension and information processing. Adaptive learning platforms, with personalized learning paths and interactive activities, help individuals with ADHD or executive functioning challenges stay engaged and focused. Virtual reality simulations or gamified learning environments offer multisensory experiences that enhance understanding and retention of complex concepts. Furthermore, online collaboration tools and note-taking applications promote organization and effective communication among peers and educators. These assistive technologies level the playing field and foster a supportive learning environment for all students.

Ethical Considerations and Challenges

While assistive technology and accessibility advancements offer tremendous benefits, ethical considerations must be taken into account to ensure responsible and inclusive use. Privacy and data security are critical concerns, as many assistive technologies rely on the collection and analysis of personal data. Safeguarding this information and providing individuals with control over their data is essential. Additionally, accessibility challenges can arise in the development and implementation of assistive technologies, requiring careful consideration of diverse needs and perspectives to prevent exclusion or marginalization. Collaborative efforts between stakeholders, including individuals with neurodivergent conditions, technology developers, educators, and policymakers are crucial in addressing ethical concerns and achieving equitable access to assistive technologies.

Future Directions and Implications

As technology continues to advance, the potential for further innovations in assistive technology and accessibility appears promising. Machine learning and artificial intelligence can contribute to the development of more sophisticated and personalized assistive technologies. Continued research and collaboration across disciplines can lead to improved understanding of neurodiversity and better tailored support systems. Inclusive design principles and standards should be integrated into technological advancements to ensure accessibility and inclusivity by default. Empowering individuals with neurodivergent conditions to participate in the design and evaluation of assistive technologies will contribute to their effectiveness and relevance. By embracing innovation and ethical considerations, society can strive towards a more neurodiversity-inclusive world, where assistive technology enhances lives and enables individuals to reach their full potential.

Summary

Innovations in assistive technology and accessibility have the potential to transform the lives of neurodivergent individuals. Augmentative and alternative communication (AAC) devices enable effective communication, while sensory support technology addresses sensory challenges. Smart home technology and environmental control systems promote independence, and health monitoring tools aid in maintaining well-being. Accessible learning tools and educational technology cater to diverse learning styles. However, ethical considerations and challenges must be carefully addressed, and future directions should focus on personalized and inclusive technologies. Through responsible and inclusive use of assistive technology, society can create a more inclusive and supportive environment for neurodivergent individuals.

Amplifying Neurodivergent Voices in Research and Policy

As we continue to embrace the neurodiversity paradigm, it is essential to amplify the voices of neurodivergent individuals in the realms of research and policy. By actively including their perspectives and experiences, we can gain valuable insights that inform the development of more informed and inclusive approaches. In this section, we will explore the importance of amplifying neurodivergent voices, discuss strategies for doing so, and highlight the positive impact such inclusion can have on research and policy.

The Importance of Neurodivergent Voices

Neurodivergent voices are crucial in shaping research and policy that directly affects their lives. Including their perspectives goes beyond tokenism; it is an acknowledgment of their expertise and unique insights into their own experiences. By elevating their voices, we enable a more accurate representation of the diversity within neurodivergent communities and empower individuals to actively participate in decisions that impact their lives.

Furthermore, including neurodivergent perspectives in research and policy helps to challenge existing biases and assumptions. Traditional research and policies often reflect the perspectives of neurotypical individuals, which can perpetuate harmful stereotypes and misconceptions about neurodivergent individuals. By amplifying neurodivergent voices, we can foster a more nuanced and accurate understanding of neurodiversity, leading to more effective interventions, support systems, and policies.

Strategies for Amplifying Neurodivergent Voices

1. Collaborative Research: Engage neurodivergent individuals as collaborators and co-researchers in studies related to their experiences. This approach ensures that their perspectives and insights are integrated into the research process from the very beginning. Researchers can work together with neurodivergent individuals to design studies, collect data, and interpret findings, promoting a more holistic and inclusive approach to research.

2. Participatory Action Research: Implement participatory action research (PAR) methodologies, which actively involve neurodivergent individuals in shaping research questions, methodologies, and recommendations. This approach recognizes neurodivergent individuals as experts in their own lives and empowers them to have an active role in the research process. PAR fosters a sense of ownership and agency among neurodivergent individuals, promoting more impactful and relevant research outcomes.

3. Community Consultation: Establish mechanisms for ongoing community consultation, where neurodivergent individuals have the opportunity to provide feedback and input on research and policy initiatives. This can be done through focus groups, surveys, or town hall meetings specifically designed to gather their perspectives. By creating a continuous feedback loop, researchers and policymakers can ensure that their work aligns with the needs and aspirations of the neurodivergent community.

4. Supportive Spaces: Create safe and supportive spaces for neurodivergent individuals to share their experiences and perspectives. Online forums, support groups, and social media platforms can serve as platforms for individuals to engage in constructive discussions, share their stories, and advocate for change. These spaces can foster a sense of belonging and empowerment, encouraging individuals to speak up and contribute their unique insights.

The Impact of Amplifying Neurodivergent Voices

Amplifying neurodivergent voices in research and policy has significant benefits for both the neurodivergent community and society as a whole.

1. Validating Experiences: By actively seeking and valuing the perspectives of neurodivergent individuals, we validate their lived experiences. This

acknowledgment helps to reduce the stigma and discrimination faced by neurodivergent individuals, promoting a more inclusive and supportive society.

2. Improved Research Outcomes: Including neurodivergent voices in research yields more accurate and relevant findings. Their unique perspectives help identify research gaps and drive the development of interventions and support systems that align with their needs. As a result, research can be more effective in addressing the challenges faced by neurodivergent individuals.

3. Informed Policy Development: Incorporating neurodivergent voices in policy development ensures that decisions are based on comprehensive and inclusive perspectives. Policies developed with input from neurodivergent individuals are more likely to consider the diverse needs and aspirations of the community, leading to more equitable and effective outcomes.

4. Empowerment and Advocacy: Amplifying neurodivergent voices in research and policy empowers individuals to become advocates for their own rights and those of others. This engagement fosters a sense of agency and promotes self-advocacy within the neurodivergent community. Through their advocacy efforts, neurodivergent individuals can bring about positive change in society.

Conclusion

Amplifying neurodivergent voices in research and policy is an essential step towards creating a more inclusive and equitable society. By prioritizing the perspectives and expertise of neurodivergent individuals, we can gain a deeper understanding of their needs, experiences, and aspirations. This inclusion has the potential to drive meaningful change in research practices and policy development, leading to improved support systems, more accurate interventions, and reduced stigma. As we move forward, let us continue to uplift and amplify the voices of neurodivergent individuals to build a more inclusive and accepting world for all.

Conclusion

Conclusion

Conclusion

In this book, we have explored the concept of neurodiversity and its significance in a neurotypical world. We have delved into the understanding of neurodevelopmental conditions such as Autism Spectrum Disorder (ASD), Attention-Deficit/Hyperactivity Disorder (ADHD), dyslexia, intellectual disabilities, and other variations. We have examined the social and cultural context of neurodiversity, discussing the history of neurodivergent communities, discrimination and stigma faced by neurodivergent individuals, and the importance of cultural perspectives in fostering acceptance and inclusion.

Furthermore, we have explored the role of education in promoting neurodiversity. We have discussed inclusive education principles and strategies, specialized programs and supports for neurodivergent learners, and the importance of teacher training and professional development. Additionally, we have examined the intersection of neurodiversity and employment, highlighting the unique strengths and challenges that neurodivergent individuals bring to the workplace. We have discussed the importance of creating inclusive hiring practices, workplace supports and policies, and the value of diversity in workforce innovation.

In considering the mental health and well-being of neurodivergent individuals, we have emphasized the significance of self-acceptance and building self-esteem. We have also explored mental health concerns that may co-occur with neurodivergent conditions, and the importance of accessible therapeutic approaches and support services. Additionally, we have highlighted the role of community support and advocacy in promoting the well-being of neurodivergent individuals.

Throughout the book, we have examined neurodiversity representation in media and popular culture, emphasizing the need for authentic portrayals and the power of storytelling in shaping perceptions. We have also celebrated neurodivergent voices in music, art, and performance, recognizing the unique creativity and innovation that arises from diverse perspectives.

Finally, we have discussed the ethical implications of neurodiversity and the need for its integration into bioethics debates, research, and policy. We have explored future directions in neurodiversity research and practice, including advancements in educational and healthcare settings, further understanding the genetic and environmental underpinnings of neurodivergence, and ongoing efforts to promote autism acceptance and inclusion.

In conclusion, this book has aimed to provide a comprehensive understanding of neurodiversity and its importance in embracing and celebrating the differences in our neurotypical world. By challenging stereotypes, advocating for acceptance and inclusion, and promoting the well-being and rights of neurodivergent individuals, we can create a society that values and benefits from the unique perspectives and strengths that neurodiversity brings. Together, let us strive towards a more inclusive future where every individual is valued, respected, and supported in their journey of self-discovery and growth.

Remember, neurodiversity is not a problem to be solved, but a celebration of the diverse human experience. Let us embrace our differences and work towards a world that recognizes and values the richness of neurodivergent minds.

% Index

Index

9 781778 904653

Printed by BoD™in Norderstedt, Germany